Build Your Own Free-to-Air (FTA) Satellite TV System

Build Your Own Free-to-Air (FTA) Satellite TV System

Dennis C. Brewer

New York Chicago San Francisco
Lisbon London Madrid Mexico City
Milan New Delhi San Juan
Seoul Singapore Sydney Toronto

The McGraw-Hill Companies

Cataloging-in-Publication Data is on file with the Library of Congress

Build Your Own Free-to-Air (FTA) Satellite TV System

1 2 3 4 5 6 7 8 9 0 QFR QFR 10 9 8 7 6 5 4 3 2 1

ISBN 978-0-07-177515-1
MHID 0-07-177515-3

Sponsoring Editor
Roger Stewart

Editorial Supervisor
Janet Walden

Project Manager
Patricia Wallenburg, TypeWriting

Acquisitions Coordinator
Joya Anthony

Copy Editor
Lisa Theobald

Proofreader
Paul Tyler

Indexer
Claire Splan

Production Supervisor
James Kussow

Composition
TypeWriting

Art Director, Cover
Jeff Weeks

Cover Designer
Jeff Weeks

This text is dedicated to my sons, Jason C. Brewer and Justin E. Brewer,
and stepsons Nickolas A. LaFountain and Phillip J. LaFountain.
It is rewarding to know that each of them is always willing
to pick up a wrench and help out with a DIY project.

About the Author

Dennis C. Brewer (Laurium, Michigan) is a Novell Certified Network Engineer and information technology solutions consultant with more than 14 years of experience in the computer technology field. He holds a Bachelor of Science degree in Business Administration from Michigan Technological University.

His interests in electrical and electronics projects date back to his middle school and high school years. He was the one who was called on in class to run the movie projectors and sound equipment and to repair the PA and alarm systems at the school. This early interest resulted in a lifelong hobby that includes collecting and repairing antique radios, TV, electronics, and kit building and troubleshooting.

Dennis enlisted in the U.S. Navy Reserve and attained the rank of Chief Interior Communication Electrician. His military service continued by completing the Army R.O.T.C. program while in college and joining the Michigan Army National Guard as a 2nd Lieutenant Combat Engineer Officer. He is a graduate of the Army Engineer Officer Basic Course at Ft. Belvoir, Virginia. For his last assignment prior to retirement as an engineer captain, he served as automation projects engineer for the Michigan State Area Command.

Dennis also spent ten years as a federal facilities management specialist with the Michigan Department of Military Affairs, where he played a key role in administering program details related to replacement, renovation, maintenance, energy conservation, and improvements of Michigan's National Guard facilities at more than 60 locations across Michigan.

He has authored numerous enterprise-level information technology and telecommunications policies, procedures, and standards currently in use by the State of Michigan, completed during 12 years as a state employee. He is also the author of *Security Controls for Sarbanes-Oxley Section 404 IT Compliance* (Wiley, 2006), *Networking Your Home or Small Office* (Course Technology, 2009), and *Green My Home!: 10 Steps to Lowering Energy Costs and Reducing Your Carbon Footprint* (Kaplan Trade, 2009); and co-author of *Wiring Your Digital Home for Dummies* (Wiley, 2006). His consulting clients have included private universities; a Washington, DC–based consulting firm; and various small businesses.

He and his wife, Penny, have two resident cats, Resa and KC. The Upper Peninsula of Michigan is currently home base for the full-time RV lifestyle they now enjoy. Plans and travel goals for Penny and Dennis and the cats include visiting every state in the lower 48 in their Fleetwood Terra LX motor home during the next five years. Dennis reports that "the roving, reading, and writing lifestyles blend together quite nicely."

Contents at a Glance

Contents

Acknowledgments

Many thanks to Mitul Chandrani of Xantrex/Schneider Electric, Brice Washington of Hauppauge Computer Works, and Steven Hutt of www.gosatellite.com for their help in getting me the pieces and parts I needed and in time and also for being sounding boards for many product-related questions.

I would also like to acknowledge the patience demonstrated and encouragement provided to me for this project by my wife, Penny. Thanks again, Penny. I love you.

Thanks also to my agent, Carole Jelen at Waterside Productions, and to Roger Stewart, the acquisitions editor at McGraw-Hill, for seeing this project as a book that needed to find its way into the DIY bookshelves everywhere.

Introduction

Satellite communication is a fascinating topic, and a person could spend a lot of quality time reading and studying all there is to know about this unique method of communication. As far as we know today, the option to watch or hear programming bounced back to Earth from a satellite in low Earth orbit began to grip both the fascination and "cold war" alarm of the Western world's citizenry with the launch of Sputnik 1 on October 4, 1957. I remember as a young lad going out into the neighbor's treeless and open yard every evening during the 22 days the satellite stayed in orbit to see if we could see Sputnik pass overhead. Not real sure if I ever saw Sputnik, but I did see a lot of bats feeding on giant insects and planes passing overhead in the process of trying to see the Russian orb. I was not aware of it at the time, but Sputnik 1 had a low power transmitter sending out an oscillating carrier wave that allowed Earth observers to track its course. I do, however, remember poignantly the push that followed in the public school systems to emphasize math and science studies for fear that without doing so, America's standing in the world and our future economic opportunities would be jeopardized. It is almost funny how the "Dick and Jane must do math" song replays in Washington, other world capitals, and in all the media today.

According to the United States Space Surveillance Network's web site, about 24,000 man-made objects have orbited the Earth since 1957. The agency currently tracks nearly 8000 objects orbiting the Earth, with about 600 of them being operational satellites. Operational satellites are still beaming back scientific readings and photos, surveillance, Earth and space physical observations, and they are the platforms providing transponders (retransmitters) to informative and entertaining programming back to Earth-bound receivers. This latter group contains the few dozen satellites that are of interest to us in this book. Of particular interest are the very few that transmit interesting unencrypted sound and video television and radio programming that can be received legally with an FTA satellite receiver without paying any fees of any kind.

The intent of this book includes a goal of breaking the "radio silence" about perfectly legal in-home viewing of FTA broadcasts. I hope to take FTA broadcast viewing from the exclusive domain of the few TV "Einsteiniacs" to mainstream use in every household.

Although counting stations in this environment is like shooting at a moving target, free to air (FTA) satellite television stations available to the United States include about 97 English language broadcast channels and more than 420 currently available channels in total, including foreign, newsfeeds, and audio/music-only channels. FTA television stations can include nationally known stations from recognizable networks such as FOX, PBS, ABC, NASA, Pentagon, and a weather forecasting channel.

FTA is rarely talked about or covered in the mainstream media and nearly unknown to the average homeowner, because the giant rival "pay-for television" providers (DISH and DIRECTTV) rule. In addition, all of the cable companies in the United States also advertise aggressively as the alternative to satellite. Tech-savvy users and hobbyists have been setting up and using FTA satellite systems since the mid 1990s. Most of the U.S. population simply but incorrectly assumes that you have to pay to get satellite TV or do something illegal to "steal" broadcasts from the pay-for providers. This book is intended to break the wall of silence on this topic and spread the word about this amazing opportunity. Even those devoted to enjoying paid satellite or cable television entertainment services can read this book and add an FTA receiver to their systems, because much of the FTA content is exclusively available via FTA channels.

At this writing, the economic downturn is changing many consumers' minds about the pay-for television economic models. Numerous articles are appearing in print and on-line about how cable subscribers and paid satellite subscribers are giving up on paying for these expensive services. Some media sources have pointed out the savings benefits of using over the air (OTA) local stations as an alternative to cable and pay-for programming over satellite. OTA is also discussed in this book for those who might not already be using this option as a program, news, and entertainment source.

For any residence or office that already has a working television set, for a one-time expense of as little as $60 and up to $600, the opportunity exists for nearly every homeowner in the United States, Canada, and parts of Mexico to purchase and set up a home satellite television receiver system. The residents of many other nations have already embraced FTA and routinely use it with programming from one or two satellite dishes attached to their homes. This book can be of benefit to readers in those countries as well, because it also treats "home television entertainment" as the overarching theme.

A system can easily combine local digital OTA stations and FTA satellite stations to receive hundreds of channels legally, including programming for music, news, television, Christian broadcasts, PBS stations, weather, foreign language television and radio, and talk radio—all for free, with no monthly charges, ever. This book will walk you through all the necessary steps to replace or simply augment your paid television services with free OTA broadcast stations using the FTA-compatible satellite equipment and components discussed in this book.

This book will attempt to keep it easy and avoid overwhelming you with the technology and jargon. The hope is that anyone can read this book and latch onto this technology and put it to use in a home or office, vacation home, or recreational vehicle.

The economic values for FTA satellite TV are also there for hobbyists who want to try it for fun or who just want to be different. The economics of this technology also work well for people who are tired of paying more to expand their TV viewing options. Setting up FTA is a onetime expense with no monthly fees—ever. There is also a case for TV junkies, who have cable and one or more satellite packages with nearly everything, to add FTA so they can take in the foreign news, movies, and other interesting programming not routinely offered on paid systems. With FTA's low and onetime cost of implementation, it makes sense to have it just to add to your viewing choices.

I have found that, on average, using the relatively new local digital broadcasts from OTA TV in metropolitan areas, between 10 and 20 channel choices are usually available. I have also traveled to areas where I have been able to receive as many as 38 channels. The story is much different in rural areas of the country, however, where citizens are fortunate to have two or three channel choices.

For those readers who do not know how to set up OTA receivers, Chapter 9 is for them. For a modest onetime expense, regardless of the population density, rural, or metro market, the added beauty of FTA satellite is that anyone anywhere in the United States with a free view of the southern sky can add this collection of channels and programming to their choice options without emptying their wallet every month.

Some channels are available only on FTA. For example, there is no incentive for paid satellite providers to include foreign broadcast choices. The Pentagon channel, for example, is broadcast to nearly every country in the world and is carried by many cable TV providers in the United States, but not by every cable TV provider. Pentagon channel programming is provided mostly by military service journalists, and it has become popular viewing for families with members serving in the military at home and overseas, retirees, and history buffs. FTA also includes a number of religious channels, talk shows, and music, with potentially broad audience appeal, if only people knew they were available. There are mainstream channels frequently viewed on paid systems such as ABC News Now, PBS, and PBS World, and some of the university channels right along with some of the more obscure channels such as Russia Today. Overall, there are enough choices broadcast in English to satisfy a diverse viewing audience.

The ideal reader of this book is anyone who operates a household (house, apartment, condo, RV motor home, and so on), owns a television and enjoys watching television, and has some dollars to spend on this book. FTA is a viable option and one of the greatest ways to save money on TV programming choices. It would be great if there were enough viewers on FTA receiver sets in the country that Oprah Winfrey's new network would create headlines by going on FTA with all her future network programming. I can see the potential headline now: "Oprah Puts *Own* Programming on an FTA Satellite Network Channel!" It would be great, but because of the money the "big boys" will pay, it probably won't happen. Actually, that is the same money that the "big boys" will charge you to see OWN (Oprah Winfrey Network) programming!

It is entirely true that even without this book, some readers could order the right pieces and parts; open the boxes; assemble, test, and set up a dish and receiver. Then hook all the connections to the TV, DVD, and 7.1 surround sound systems and it would work the first time. However, some others would be fearful of even attempting such a task. In fact, many people are challenged simply from moving a DIRECTV or DISH receiver box out to their RV and aiming the antenna when they get to a new location. There is a lot to know, and this book presents information helpful to those using and moving the paid services to multiple locations as well. My hope is that whether you are a plumber, CPA, or hair stylist, you could read this book and have the confidence to order and install all the components of an FTA system. It is not rocket science, but it is technical enough that this book can help any DIY person through the process to success.

As you browse and read this book, all chapters after Chapter 1 can probably be read in any order, except it might make it easier if you read Chapters 3–7 in sequence. These five chapters have a primary focus of presenting the information most needed for setting up your FTA system.

Chapter 2 will help you get started with selecting the tools you will need to match the scope of your project, so reading Chapter 2 just before you start work is fine. Chapter 9 is an important chapter for getting "local" stations from OTA to your TV screen. For those thinking about new, upgrades, or adding TV sets Chapter 10 is helpful for firming up any purchasing plans you have. Chapters 11–14 are all about increasing the quality, available variety, and quantity of options for your home TV and entertainment viewing experience. For those who want to leverage the power of their PC and use their computer as the muscle for TV entertainment as a media center, Chapter 13 should provide enough information to get you started on a project of installing a TV card in a PC. If you are really fanatical, for whatever reason, about home entertainment, Chapter 15 will provide a few hints about dedicating a room to use exclusively as an in-home theater. And, finally, for the readers who can't wait to attend the next all-day tailgate party or travel to a favorite campsite for the weekend, Chapter 16 will provide information about "going rove" with your FTA TV.

With high unemployment rates and families facing inflated prices for basics such as food and gasoline, everyone, even the affluent, are looking for new ways to save a few dollars on a regular basis. At the current rate of about $70 per month or more for decent plans on DISH, DIRECTV, or cable fees, installing FTA could save you about $2500 over three years. FTA, although free of monthly bills, still presents good choices for TV programming for any average homeowner with a few tools, $100 to spend, and some of the information in this book.

In America, as citizens, we think of the airwaves as belonging to us, all of us, and in a sense they do. An FTA system installed as a result of this book can bring hitherto unreachable programming into nearly every home in the country. I am sensing the potential for a paradigm shift nearly as big as some of the changes brought about by the Internet once the information barrier about the existence of FTA and the relative ease of installing FTA is smashed by the wide distribution and reading of this book.

I hope that you find this an interesting read and in so doing find the confidence to make your FTA project a complete success for yourself and your TV viewing companions.

CHAPTER 1

Navigating Your Way Through the Jargon Jungle

A whole new foreign language seems to spring up right along with any new technology these days. It truly is a jungle out there when it comes to understanding jargon and the countless acronyms and cryptic messages with which we are bombarded on a daily basis from the various aspects of the professional, political, technical, and socially connected world in which we live. Regrettably, there is no worldwide arbitrator and keeper of jargon, tech-speak, or slang terms.

To a government agency, the acronym *FTA* could refer to the Federal Transit Authority in Washington, DC. If you are listening to a political speech on the Capitol steps, FTA might mean Free Trade Agreement. When you were in high school, FTA probably meant Future Teachers Association. To someone in the legal profession, FTA is a citation for failure to appear in court. In this book, of course, *FTA* stands for *free-to-air*.

With the meaning of words changing based on time, culture, location, context, and subject matter, we find it increasingly important to capture, understand, and chart our ways through the lingo associated with setting up a free-to-air satellite system. You won't get quizzed on these later, but you do need to learn and embrace many of these terms, if only during the time you are purchasing and setting up your own FTA system. For those of you who forget, you can always refer back to this chapter for a refresher if you need a reminder of their meanings. Once your system is set up and running the way you want, about a dozen jargon words applying to FTA satellite reception will become resident in your argot.

The opportunity to perceive meaning in hearing tech-speak that differs from reading tech-text is that in the reading, you usually have time to use a reference source to find and study the definitions of unfamiliar terms. People who rattle off jargon and tech terms while speaking are often annoyed by questions, and novices to a topic are often embarrassed to stop their learned colleague's dissertations and ask for definitions or clarification. When failure to question happens, however, learning opportunities are needlessly lost. When you hear tech-speak, be emboldened to ask questions when you have them, and you will either learn the expressions' meanings or learn that the expressions have no meaning for the talker either. In either case, you will know more after you ask.

Simple definitions can help you understand completely when they include the necessary technical points. On that note, it is my hope the definitions here are sufficient for your needs. If you desire to delve into the finer points and nuances of these terms, you'll find many technical books and technical dictionaries out there to provide additional understanding. To save money, check to see if your local library has a reference section that contains information on the topics you are interested in learning more about.

NOTE *The definitions in this chapter and those throughout the book are used by the author in the context of their relevance to FTA satellite installations. Definitions for nontechnical words and terms used in the book can be found in other McGraw-Hill texts such as Everyday* American English Dictionary *or* McGraw-Hill's Dictionary of American Slang and Colloquial Expressions.

When I come across a term, abbreviation, or acronym I do not understand, I find it best to stop reading and focus my energies and time on finding the definition that fits the term's usage and context. Then I go back and start reading anew with the definition in my grasp. Presume and you presuppose; if you are wrong you must resuppose. The best navigation tip for finding your way to understanding through new technical information is to presuppose nothing. When in a quandary over a word's meaning, find out to calm out.

The terms presented in this chapter are a primer of those used potentially throughout the book and in other references and resources for FTA satellite TV, home theater, and audio system information. Other terms used and limited to the topics within the other chapters are defined within those chapters or are hopefully easily understood within the context of those chapters.

A

A.C. Alternating current electricity, which alternates directions as it flow though a wire.

A.K.A. Shorthand for also known as or is an alias.

AM Amplitude modulation (AM) is a method of transmitting signal information where the frequency or carrier does not change and the information (voice, music, or data) is encoded by varying the strength (amplitude) of the signal. Most automobile radios have an AM band for listening to talk radio, music, and messages from the Emergency Alert System (EAS) in the United States.

Audio Sound carried by or reproduced from something other than its natural source.

C

C Band Satellite frequencies above 3.9 GHz and below 6.2 GHz.

CAT 5 Category 5 computer network cable comes in two types: straight-through for connecting unlike devices and crossover for connecting like devices. Some network connections/cards will perform automatic crossover when needed.

D

dB Decibel; a unit of measure used to quantify the intensity of sound, expressed as a ratio exponential of a logarithmic function: *10 × log(power out/power in)*. It can be applied to compare any force such as sound, electrical power, radio waves, or light. It is used to compare gains or losses that are large, where a linear comparison would be less descriptive or more difficult to express. dB is used, for example, to show the signal degradation when a tower emits a megawatt and the receiver's antenna picks up a thousandth of a watt. Without using dB, that difference, either gains or losses, would be difficult to compare.

D.C. Direct current electricity, which flows one way through the wire.

DiSEqC Digital Satellite Equipment Control refers to protocols for controlling switches and motors.

DIY Do-it-yourself.

DTV Digital television; includes the features of multicasting and data streaming. It is the method of signal transmission that makes *HDTV* possible.

DVD Digital video disk also used to refer to a digital video disk player.

DVR Digital video recorder, a device which stores video and audio files and plays them back when selected.

F

FCC Federal Communications Commission is an agency that regulates all forms of electronic communication in the United States and cooperates with the United Nations and with other countries to maintain the usefulness of worldwide communications media and frequency bands for everyone.

Firmware A software program or operation system embedded into silicone chips.

FM Frequency modulation is a method of transmitting signal information wherein the frequency or carrier changes and the information (voice, music, or data) is encoded by varying the frequency of the carrier signal. Most automobile radios have an FM band for listening to talk radio, music, and messages from the Emergency Alert System (EAS) in the United States.

FTA Free-to-air; FTA satellite digital television programs are sent in the clear without encryption so that any satellite tuner can receive and decode the signal into video, audio, or both.

G

GFIC A.K.A. GFI; an outlet or a circuit breaker equipped with a ground fault interrupter circuit intended to reduce or eliminate lethal electrical shocks.

GHz Gigahertz; 1 billion Hz, or 1 billion cycles per second.

H

HD video High definition video.

HDMI High definition multimedia interface; a standard for connecting video/audio devices to other devices and TV sets.

HDTV High definition television; offers superior picture quality, surround sound, and a movie theater–like wide-screen format.

Hz Hertz; one cycle per second.

K

KHz Kilohertz; 1000 hertz, or 1000 cycles per second.

Ku Band K-under band; for FTA satellite, frequencies above 10 GHz and below 13 GHz.

L

L Band The 950 MHz to 1750 MHz frequency range on the down-lead cable from the LNB to the receiver. These are the frequencies the FTA receiver can process and decode.

LNB or LNBF Low noise block or low noise block feed. Three signal polarities are possible for the LNB/LNBF: vertical, horizontal, or circular. In this text, LNB is used for both LNB/LNBF. As the focus is FTA where a horizontal and/or vertical polarity LNB is needed. Circular polarity is needed primarily for the pay-for services on satellite. The role of the LNB is to convert the higher frequency satellite signals to lower band signals that the receiver can process and decode and that the cable can carry successfully from the dish to the receiver. See *L Band*.

M

Megabit (Mbit) When used in computer or communications technology a bit is a one or a zero, a megabit is one million bits.

Megabytes When used in computer technology a byte is equal to eight bits, a megabyte is a million bytes.

MHz One million hertz, or a million cycles per second.

O

OS (operating system) A program that runs the computer's hardware or runs a computer-like device.

OTA television Over-the-air television is received over an indoor or outdoor antenna.

P

PC Personal computer, notebook or laptop.

PVR Personal video recorder; records video and audio for later use.

R

RG6 An insulated shielded coaxial cable for use on high-frequency and digital applications that is commonly used for satellite down leads and cable frequency applications.

RV Recreational vehicle; a class A, B, B+, C motor home, travel trailer, van, or caravan.

S

Skew Tilting the low noise block (LNB) angle to match the horizon of the satellite transponder's signal polarization. Not necessary for circular polarized signals but very necessary for FTA signals.

Skype A trademarked software program that allows a person to make voice and video phone calls over the Internet inexpensively or for free. This software program has been ported to work on computers, handhelds, cell phones, and many other devices that are capable of communicating over the Internet.

Software A program that runs on a computer's operating system.

SUV Sport utility vehicle; originally used to refer to multipurpose vehicles built more like station wagons on a beefed-up chassis. An SUV is great for carrying portable FTA equipment to and from a tailgate party.

U

UHF Ultra high frequency; in TV context, it refers to over the air channels 2 through 13.

USB Universal serial bus; the standard for connecting computer peripherals to a computer or portable drives to a computer-like device.

V

VCR Video cassette recorder.

VHF Very high frequency; in TV context, refers to over the air channels 14 to 69.

VHS Video home system; a tape cartridge and format used for recoding with a VCR.

Video Images or moving pictures carried by or reproduced on digital tape; can include the associated sound.

W

Watt Calculated rate of electric power derived by multiplying amperage times voltage.

Consider yourself now equipped with enough jargon and technical terms to tackle the rest of this book. Enjoy the read, take a few notes along the way, and begin the rest of your journey to technical competence with the FTA language.

CHAPTER 2

Tools and Equipment

First of all, do not let this chapter on tools and equipment frighten you away from buying and installing a simple or a first FTA system in your home, cottage, apartment, or RV. For a simple system, everything you might need can be found "off the shelf," and the most important tool you will need is simply some measure of determination to plan the project and get it done.

This chapter is for those of you who want to do it all—to go beyond the simple, build more than one, or build one for mom or a special friend as well as your own. Or perhaps you think building FTA systems might make a good "moonlighting" job to make a few extra dollars on weekends. But even if you are putting together the simplest system, you will need a tool or two, so every reader can benefit from this chapter to learn when a particular tool will be helpful for completing the project. You do not need to own every tool mentioned here to get the job done, but you will likely need some of them.

Do-it-yourself (DIY) projects proceed with relative ease when you combine knowledge of what needs to be done with the right tools to get the job done. A small number of essential handyman tools will carry you through most of your FTA satellite system and TV-related installation projects. A few specialized tools are helpful in maintaining the project momentum and will also help improve the overall quality and craftsmanship of the final installation. Some specialized tools can also reduce the installation costs by allowing you to use, for example, bulk cable that you cut to length, instead of preassembled cables in blister packs already cut to length with end connectors already attached.

Sometimes the appropriate pieces of electrically powered tools can be helpful in saving time and effort. For example, when a concrete block wall or large beam must give way to a hole for the satellite dish's cable or a mounting anchor, you could consider renting a heavy-duty hammer drill and heavy-duty masonry bits if you don't have them.

If you are starting this as your first DIY project and you have no tools, yet you want to maintain this as a low budget project, consider borrowing needed tools from a friend or neighbor. Otherwise, consider this FTA project as a reason to begin to collect some tools that will work for this and many other DIY projects.

Safety First

A word of caution is in order for any reader who will be using hand and power tools. No matter what the job or project, the most important tool to put in your "virtual" toolbox is a safety state of mind.

Never rush a job that has any potential for accidents. Take your time and think the project through before you begin to determine what safety equipment you should have and use. Then find the safety equipment and list the safety precautions needed before you begin and use them at the appropriate steps along the way. For example, always wear safety glasses when drilling or hammering, and wear a hard hat and steel-toed shoes or boots when the risks warrant using them.

Have a safety person present when using ladders or overhead lifts to help keep you safe while you work, and have them wear a hard hat. Have an OSHA-approved dust mask when necessary to protect your lungs from airborne particulates and things like the dust particles from fiberglass or mineral insulation.

Remember Smokey Bear's statement that "Only you can prevent forest fires"? Well, it's the same with on-the-project accidents: "Only you and your fellow workers can prevent accidents or injury on the job."

Installing an FTA system will not require you to go out and break the bank by buying tools. Compared to many other DIY projects that are tool-intensive, FTA installations are fairly simple projects that can be accomplished successfully with a fairly low tool count.

If you are new to DIY projects, the following tool introductions might prove useful when you proceed to the installation phases of your project. If you are experienced with DIY and have a substantial tool chest already, you might simply want to use this section as a checklist to ensure your tool treasure trove includes items like these.

To round out your tool collection for this and future projects on a tight budget, consider shopping at thrift shops, garage sales, rummage sales, and flea markets. Armed with a list of what you need, coupled with a careful eye to look them over and a few dollars in your pocket, you can often find very serviceable tools on the cheap at any of these places. A major theme of this book is about free programming, so it is fair to mention that if finding tools can't be free, it's cousin, buying cheap, is OK, too.

To assist you with building the tools needed list for this project I will also suggest FTA project elements for which the tool mentioned might be useful.

Ordinary Hand Tools

Ordinary handyperson tools are those that are intended to be used for multiple purposes, or one could say general, as in intended to be used for no specific single purpose. Simple hand tools might not always be the best choice for a job, but they are usually inexpensive to buy and are often already found in unusual places in the home such as in the "junk drawer" or hiding under a kitchen sink.

Before you buy new tools for any for your FTA projects, it is a good idea to take inventory of what you already have. Once your tools are collected in one place, take a few more moments to inspect them for serviceability. Phillips-head screwdrivers aren't much good if the blade tips are rounded over or gouged out. Hammers with loose handles become much too dangerous to use safely.

After you have identified the junk tools, find or buy a nice toolbox, or at least select a single secure place to keep the good tools and send the junk one off to the recycling center.

CAUTION *After a moment of silence for their years of service or a few bars on the trumpet if you must, please discard the unsafe and no longer useful tools or electric powered equipment as another necessary precaution to prevent future injury to you or anyone in your household who uses the tools.*

Large Pliers

Figure 2-1 shows four types of large pliers. Clockwise from left to right we have adjustable pliers, sometimes called tongue and groove pliers. At top-middle are adjustable locking pliers with a C clamp handle that allows it to double as a small vice to hold small work pieces in its jaws when the clamp is attached to any solid location or workbench. At top-right are regular pliers, often called slip-joint pliers; at the bottom center are straight jaw locking pliers

The common use characteristics of this group of tools are the speed at which they adjust to hold something and quickly provide leverage to tighten, loosen, or bend something. On your FTA project, pliers can prove helpful for such things as assisting with tightening the mounting hardware for the satellite dish if you do not have the exact size wrenches or sockets. Or pliers can help loosen a tight cable connector.

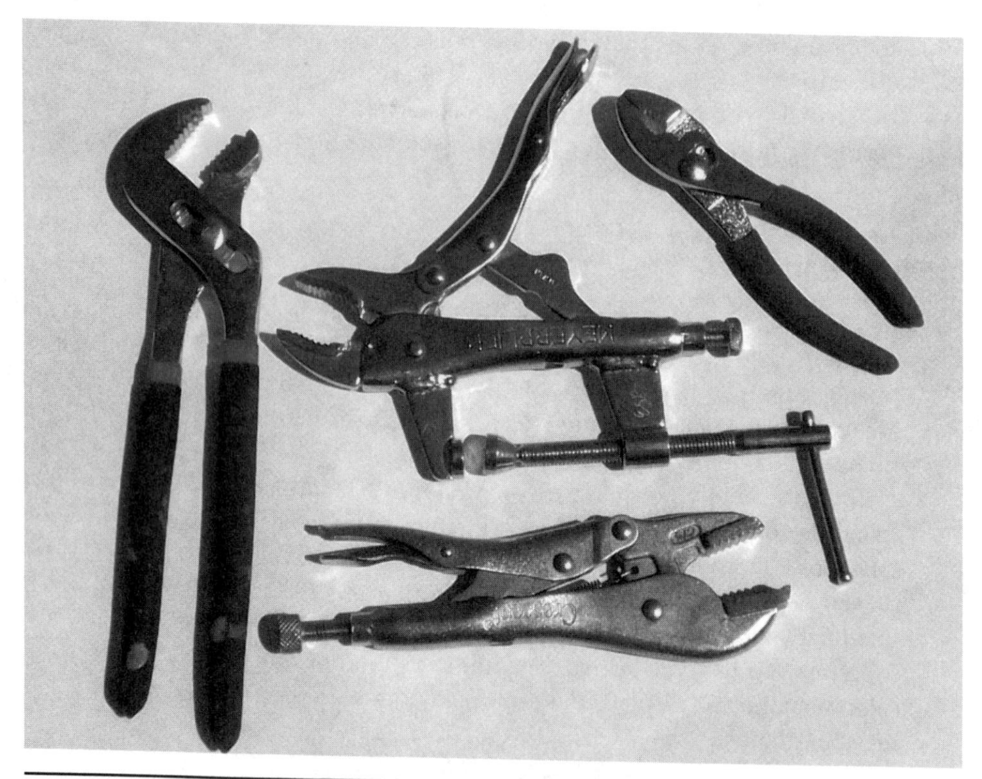

Figure 2-1 Large pliers: (clockwise left to right) Adjustable tongue and groove pliers, adjustable locking pliers with a C clamp handle, slip-joint pliers, and straight jaw locking pliers

Small Pliers

Small pliers, shown in Figure 2-2, (counterclockwise, starting from 7:00) include needle nose (long), diagonal cutting pliers, needle nose angled, nippers, duck billed pliers, and needle nose (short). The nippers are handy for pulling out small nails or screws with broken heads. Needle-nose pliers are great for reaching into tight spaces to grip something or for holding small parts or bending wire. The diagonal wire cutting pliers are designed for and are best used for cutting wire such as speaker wires to length.

Screw, Nut, and Bolt Drivers

Various sizes and types of traditional screwdrivers form a circle in Figure 2-3. Flat-blade screwdrivers and Phillips-head are the most common types and come with the drive tips metered to different sizes. Flat-blade screwdrivers vary by the width and the thickness of the blade to fit various sizes of single slotted or cross-slotted screw heads. Phillips-head screwdrivers come in five sizes numbered 0 to 4 to match the various head sizes of the different Phillips-head screws.

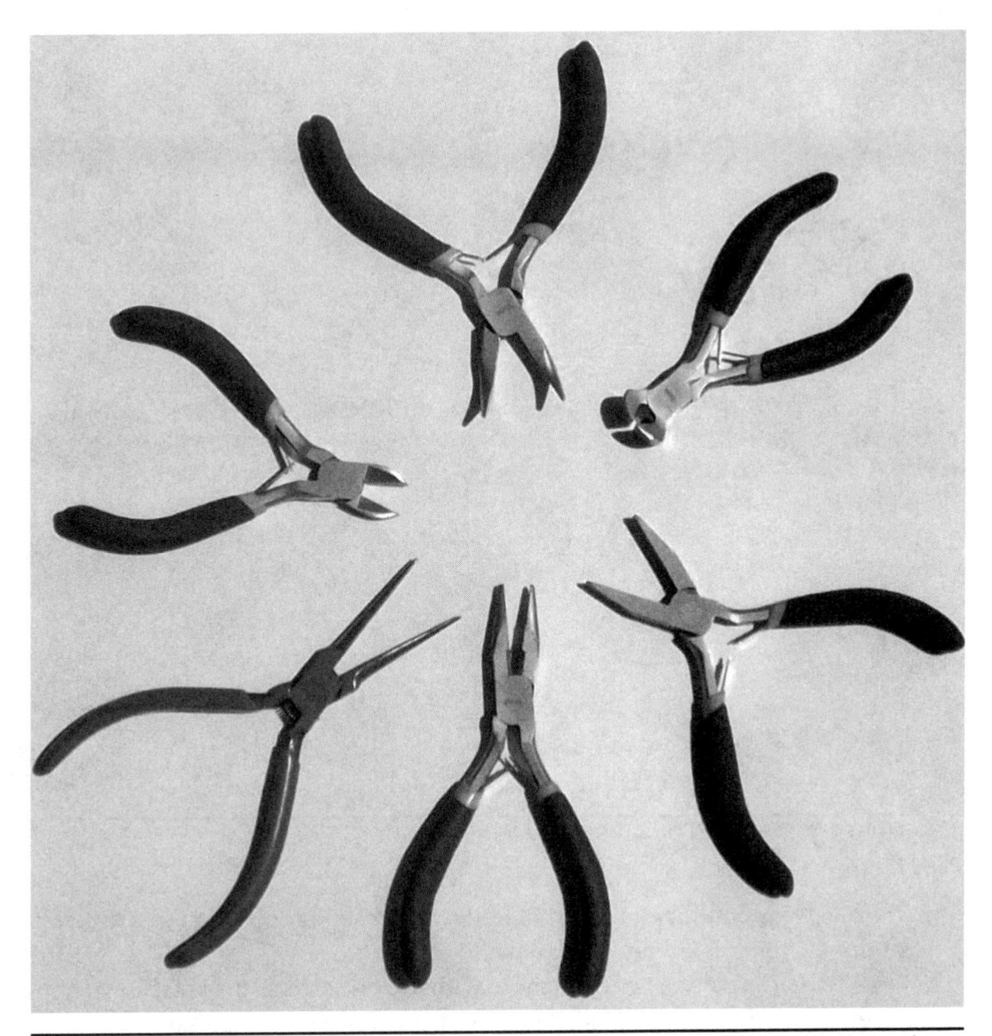

FIGURE 2-2 Small pliers (counterclockwise, starting from 7:00): needle nose (long), diagonal cutting pliers, needle nose angled, nippers, duck billed pliers, and needle nose (short)

There are many possible screw head types, such as an indented square head, hex, and many other rare specialty styles. The variability of screwdrivers carries beyond the blade sizes to include different length shafts and various handle types. You might need more than one type of screwdriver on your FTA project for tightening mounting screws, fastening connections, or opening an access port. You have to match the screwdriver to the screws you are using, so check things over before you start to make sure you have the driver type you need.

Figure 2-4 shows a simple alternative to owning a lot of traditional screwdrivers. The shaft on this multibit screwdriver allows you to insert various types and sizes of screwdriver "bits" into the shaft to match various screw heads. When

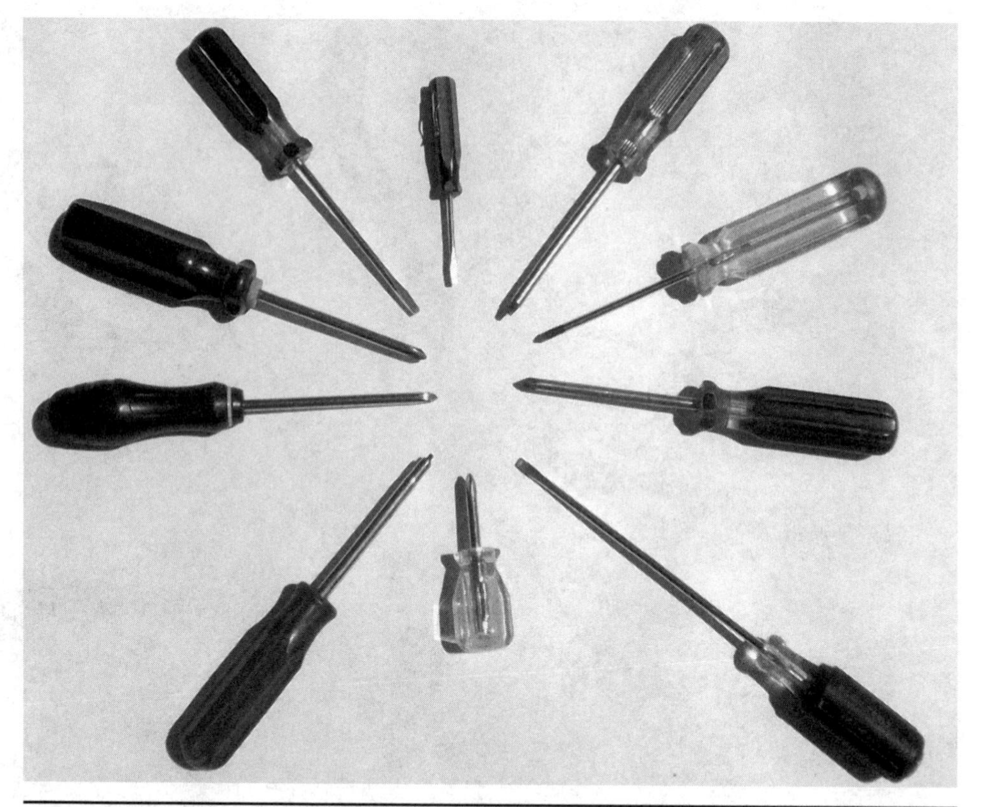

Figure 2-3 Traditional screwdrivers

the bits wear out, they can be replaced, and when you need to use a different type of driver, you can simply buy a new bit.

You can also use a ratchet handle with the bit, such as the one shown above the screwdriver, to increase the leverage to make it easier to tighten or loosen a stubborn screw. The value of using bits can be enhanced by using an adapter (at an angle and to the right of the screwdriver in Figure 2-3) that fits in a drill chuck or battery-powered driver that magnetically holds replaceable bits in its shaft.

Below the screwdriver is a small assortment of bits in a plastic holder, and additional bits are stored in the handle of the screwdriver. Although the multibit screwdriver does not let you work with very large screws, it is still a good alternative to have in your tool box.

Bigger screwdrivers are often inappropriately used as pry bars, so if you have a proper pry bar, you might not miss the big screwdrivers.

Bit-style drivers will work for most of the screwdriving needed for a typical FTA installation. It is also worth noting that you can buy other bit styles including nut driver bits, small drill bits, and socket adapters to fit the shafts and adapters. Most hardware stores will sell the bits one-off or in dual packs so you can buy them

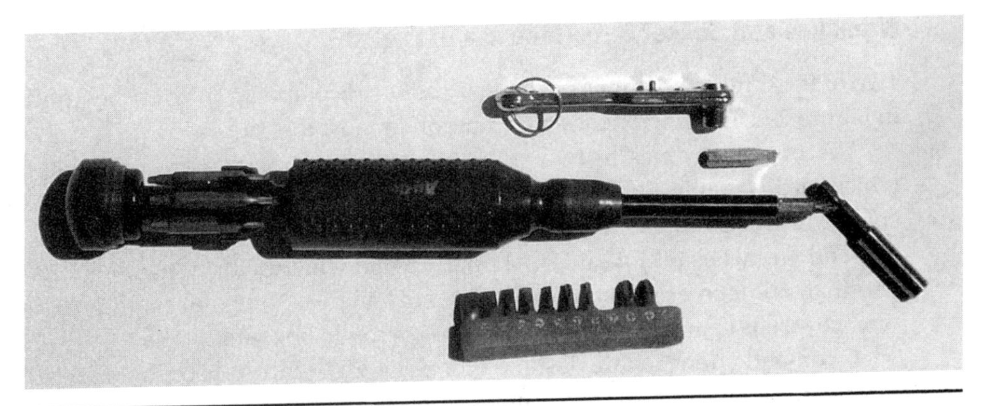

FIGURE 2-4 Multibit or replaceable bit screwdrivers

as you and your projects require them, or you can treat yourself to a larger prepackaged kit that includes nearly every kind of popular bit.

A set of small hex key wrenches are show in Figure 2-5. Hex keys or Allen wrenches are available in both SAE/English measure and metric sizes. Those shown in the figure are in fractions of an inch and are measured by the distance across any two parallel sides of the hex. Hex key screw heads and bolts are often found on assemble-yourself furniture such as TV entertainment centers. They can also be found included in mounting hardware and are often used as set screw heads to hold analog dials on sound equipment or receivers, because the hex keyhole can be nearly as small as the diameter of the screw.

FIGURE 2-5 Small hex key wrench set

Wrenches and Socket Sets (Metric and English)

Two of the same size ($\frac{1}{2}$ inch) open-end/box-end combination wrenches are shown in Figure 2-6. These wrenches are common in most tool boxes.

The box-end comes in two styles. The box-ends shown are called six-point, which means that in tight spaces slightly more than 60 degrees of rotational throw space must be available for the handle when turning a hex nut. You can also buy box-end wrenches with 12 internal points that are handy for tightening nuts when less than 60 degrees of rotational space is available. Notice in the figure that the wrench size is stamped into the handle close to the box-end.

Open-end wrenches are designed for use with square nuts or square head bolts. Open-end wrenches also work on hex nuts and bolts, but caution is in order not to round over the points on the nut or bolt. When you need to apply a lot of torque, do not use the open-end on hex nuts and bolts. You will likely need wrenches similar to these to tighten mounting hardware for the dish antenna or for assembling a mounting tripod.

Figure 2-7 shows a wrench similar to those in Figure 2-6, but this wrench is in a metric size. Notice the *14* stamped in the handle, for 14 mm (millimeters). When assembly is necessary, you might need to use metric-sized wrenches, because the products you buy can originate in any part of the world. Check the assembly instructions and look for a heading for *tools required* in the book or service manual that comes with the product to determine whether you need metric sizes for the assembly tools.

Wrenches are also available with distinct features, such as the one shown in Figure 2-8. This wrench has an offset handle, with a different size on each end—in this case, $\frac{1}{2}$ inch on one end and $\frac{9}{16}$ inch on the other. When working on older bicycles, a person needed this wrench and a similar one with $\frac{7}{16}$ inch on one end and $\frac{3}{8}$ inch on the other end for most maintenance and repairs. The big benefit of this style of wrench is simply that fewer tools are needed in the tool box. This style

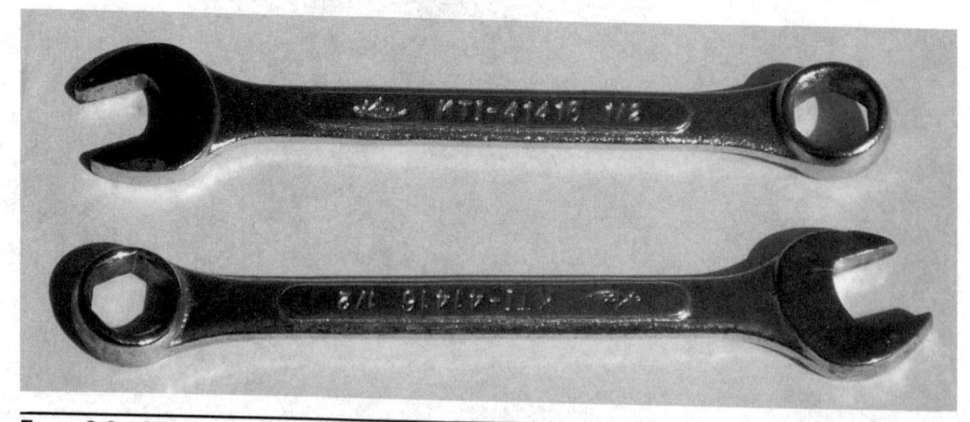

FIGURE 2-6 Open-end/box-end combination wrenches

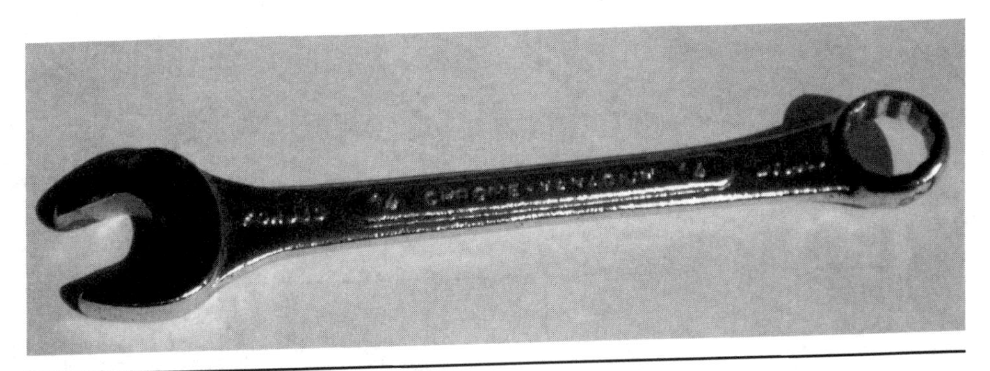

Figure 2-7 Metric wrenches

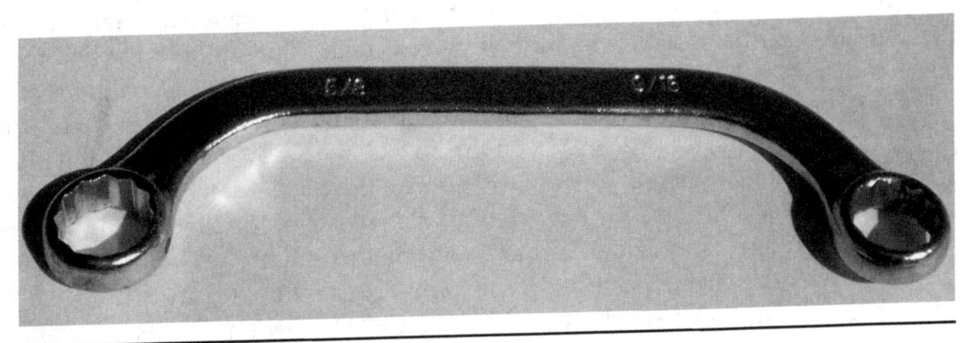

Figure 2-8 Offset box-end wrenches

will also work for assembling dish-mounting hardware. Sometimes the offset helps in limited working spaces.

One more unique wrench is shown in Figure 2-9. One of the sides of the hex is cut away. These wrenches were designed for use with fittings on copper and other types of tubing, as the slot in the hex allows for the wrench to be placed over the tubing and slid up to the fitting for tightening or loosening. They can be used like any other box-end wrench. This set also has a different size on each end and would work to loosen or tighten mounting hardware nuts and bolts. If you are doing a lot of F-style connectors, a cut-out wrench, or tubing wrench, like this can be useful.

Two ratchet drive handles are shown in Figure 2-10, one a ½-inch drive and the smaller handle a ⅜-inch drive. The size refers to the size of the square that is pressed into the socket, allowing the wrench handle to provide leverage. The ratchet handle works to provide rotational torque in one rotational direction or the other based on the setting on top of the drive. This feature allows the socket to stay in place and the handle can be rotated back and forth while turning the nut in only one direction. This feature is also very useful in tight spaces.

When working with mounting hardware that has nuts and bolts, keep in mind that the best method is simply to place the bolt through the appropriate holes and

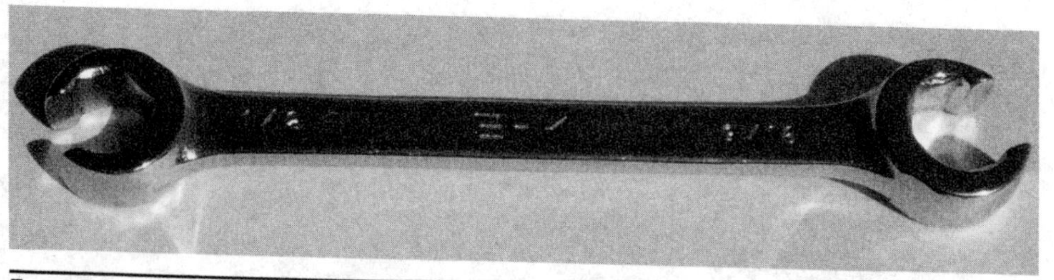

Figure 2-9 Tubing wrench

to use a box-end wrench with one hand to hold the bolt stationary. With the bolt held firmly in place, put on any washers or lock washers required and use the correct sized and appropriate depth socket to rotate the nut tight with a ratchet wrench in the other hand. If this is difficult, recruit an assistant to hold the bolt for you. With most bolt threads, that means rotate to the right (clockwise) to tighten and left (counterclockwise) to loosen. Turning the bolt and holding the nut is not correct, because it damages the threads on the bolt as it "chews" at the mounting flanges and mounting material. The short handle on the smaller ratchet wrench shown in the figure is a favorite for light-duty work because of the short length; it is quick and easy to work with and fits nicely into smaller spaces.

Two different size ½-inch drive sockets are shown in Figure 2-11. These fit the larger size ratchet drive shown in Figure 2-10. The taller one is referred to as a deep socket and its purpose is to tighten nuts on bolts with a longer throw. Notice that, just like the wrenches, the size of the socket is stamped on the side. Sockets are

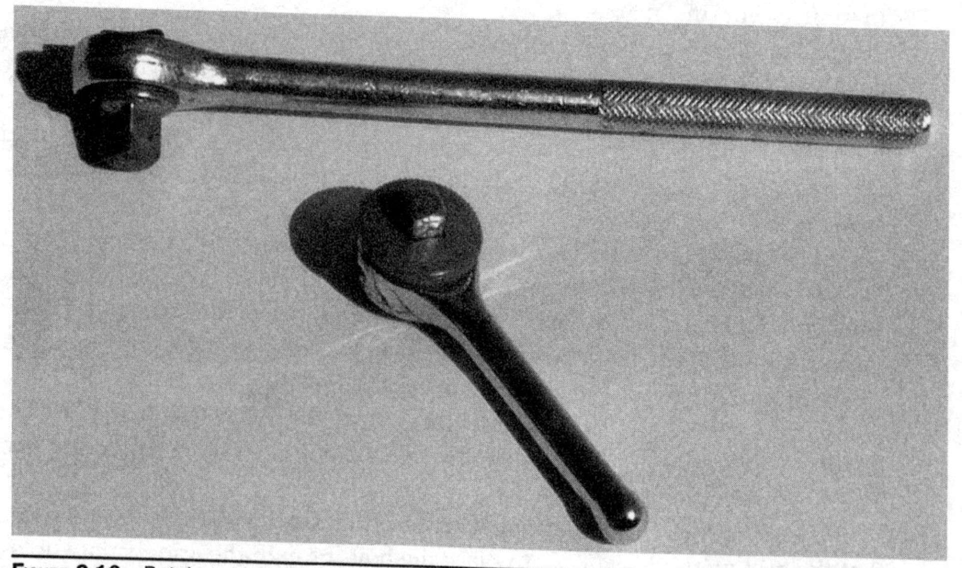

Figure 2-10 Ratchet drive handles

Figure 2-11 $\frac{3}{4}$ inch and $\frac{13}{16}$ inch, $\frac{1}{2}$-inch drive sockets

readily available in $\frac{1}{4}$-, $\frac{3}{8}$-, and $\frac{1}{2}$-inch drive sizes in most local hardware stores. Sockets are available in all standard production sizes for the SAE (English) or metric sizes. Specialized sockets with hex keys and other specialized drives are also available.

Figure 2-12 illustrates a metric deep socket and shows the *19mm* stamping denoting that the socket is intended to fit a 19 mm nut or bolt head.

Figure 2-12 Metric deep socket

Drills, Saws, and Cutters

Handyperson projects, including the occasional FTA installation, frequently need some part involved in the project to be modified, custom-made, or adapted in some way to fit or fit better. The tools presented in the next few figures all share the purpose of modifying something. These tools might be needed outside to make mounting holes for a satellite mounting bracket or inside the home to alter an entertainment cabinet.

A standard electric-powered drill motor with a $\frac{3}{8}$-inch chuck is shown in Figure 2-13. The drill chuck will fit drill-bit shanks up to its size. In the drill's chuck is a hole-saw used for cutting round holes to a given size from a pilot center point. A spare saw component is shown nearby. A hole-saw is composed of the arbor, which serves three purposes: Its stem fits in the drill's chuck to transfer the turning force from the drill motor. It holds the pilot drill bit (usually with a set screw) so the hole will be centered on the pilot point. Finally, most hole-saw arbors are threaded to hold different sizes of the circular-shaped sawing blades. Complete hole-saws can be purchased separately for the sizes you need or in a kit containing many different sizes of changeable blades. Hole-saws are very useful for cutting perfect circular holes in composite pressboard, plywood, softwood, and hardwood.

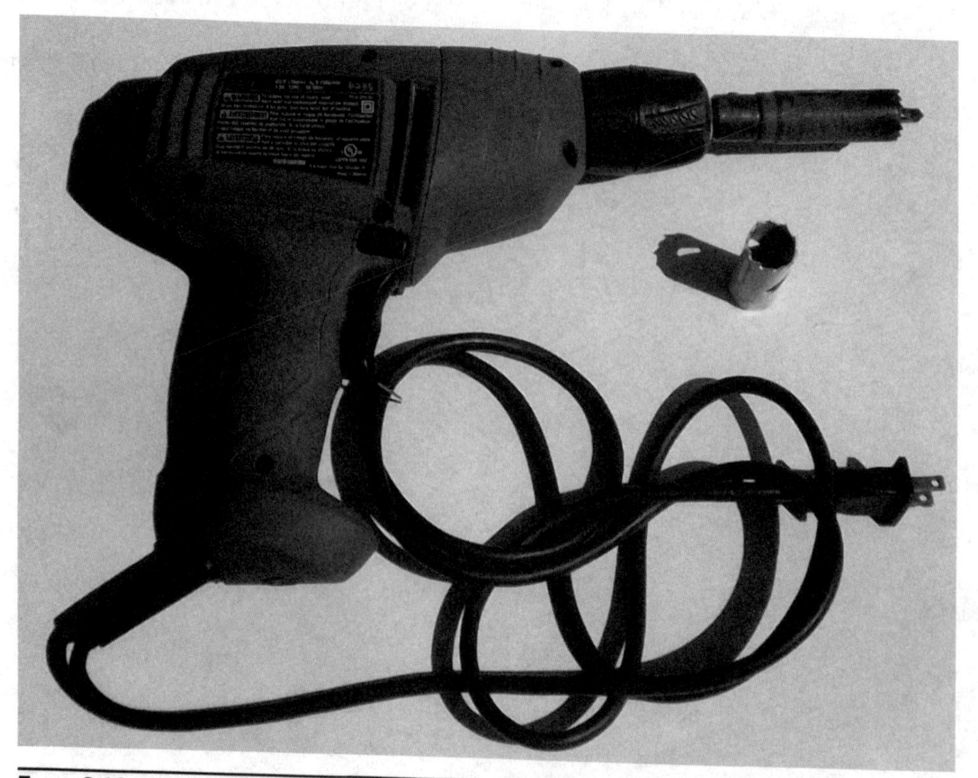

FIGURE 2-13 Drill motor with hole-saw in chuck

Many very nice battery-powered drill motors are available, and nearly all of them are sufficient for DIY projects. I have a bias for using corded power tools because I like to count on the torque being available and do not like running out of power during the project. A second advantage to corded power tools is they tend to be less expensive; so if you're buying a drill motor for the first time, considered going corded. The downside of corded power tools is that you have to keep your virtual "safety cap" on and be acutely aware of the cord and its location while you work.

Figure 2-14 shows a larger hole-saw and the cutout piece it made from a pressboard desk.

Standard drill bits are shown in Figure 2-15. If you will be mounting your satellite dish to a building, you will probably use standard drill bits for pilot holes for the screws that hold the mounting brackets to the structure or for holes to thread the satellite dish down-lead through as it snakes its way from the dish to your receiver. Again, you can buy drill bits one-off or in a set in metric or English sizes. For drilling through masonry, you need a specialized masonry bit, and you might also need a hammer drill motor to press a hole through thick masonry, stucco, brick, or concrete wall material.

Figure 2-16 shows two extended-shaft drill bits. The bottom bit has a standard twist bit on the end, and the top one is a paddle bit, so named because it looks like a paddle or oar for propelling a boat. Paddle bits work well for drilling holes in studs and floor joists through which you can run wires and cables.

You could also use a drill bit extension shafts, and those work well as long as the extension shaft is smaller than the size hole you are trying to drill. These long-shafted drill bits work nicely for drilling completely through a 2-by-4, 2-by-6, or even a 2-by-8 stud construction wall. You might need to do that once or twice as you thread your satellite dish down-lead from outdoors to its destination location

FIGURE 2-14 Hole-saw and cutout piece

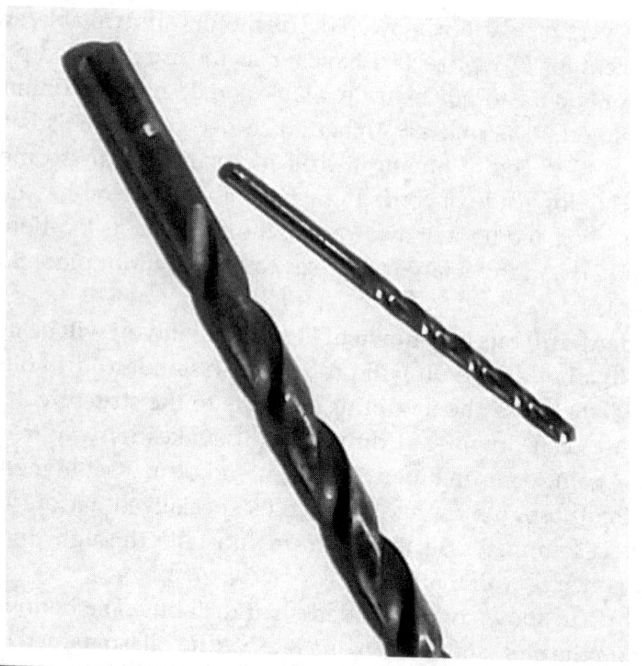

FIGURE 2-15 Standard drill bits

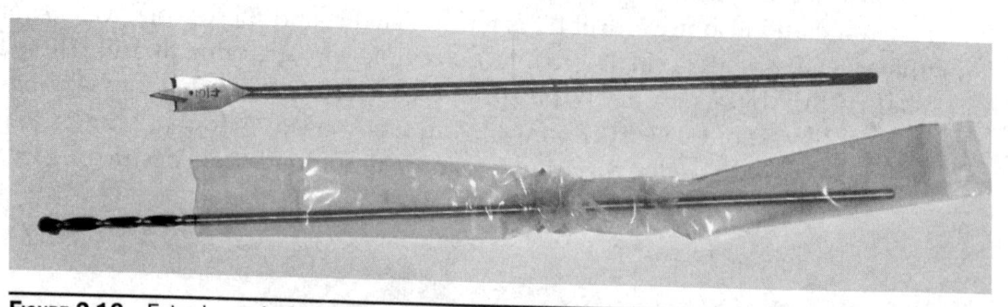

FIGURE 2-16 Extra-long shaft drill bits

inside next to your satellite receiver location. Perhaps you will also use these bits again for interior wiring for other video cable or audio wire runs.

CAUTION *Always be sure that water pipes, drain pipes, electrical, or other communication wires are not present in the wall cavities before you begin drilling through. Your safety cap should include the old adage, "check twice, drill once."*

The handsaw shown in Figure 2-17 is commonly referred to as a hole-saw, too, but so we don't confuse it with the drilling hole-saw, we will drop the shorthand and call it by a more proper name: keyhole saw. Should your plan for wiring include installing wall outlets for video/TV cable, network cable, or sound system

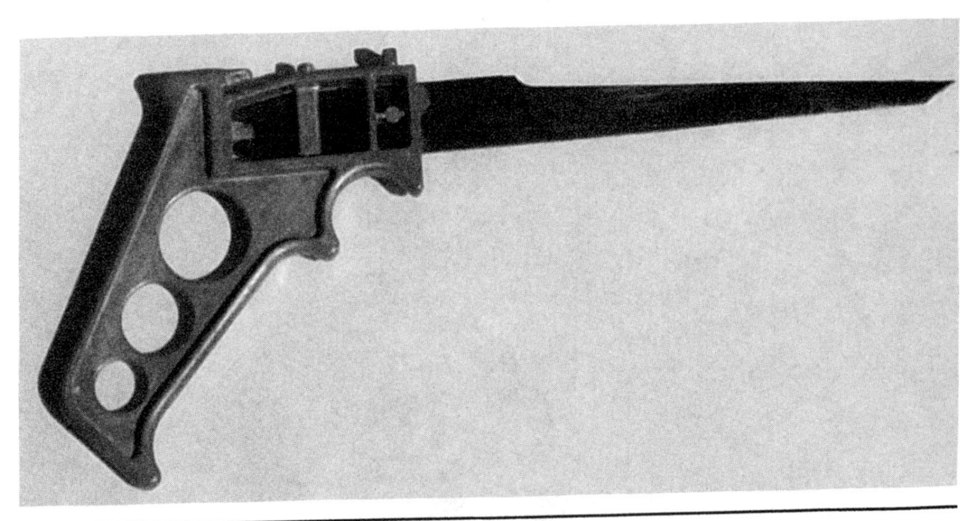

FIGURE 2-17 Keyhole saw with changeable blade feature

speakers, a keyhole saw will be used for cutting the outlets into the paneling, drywall, or other wall material. When you need to drill a hole for wall outlets, first turn the outlet box over, trace its outline where you want the box on the wall, and then drill in the "keyhole" near a corner of outline; this is where you'll place the tip of the keyhole saw's blade the into the wall and begin sawing along the lines until the cutout is complete. As always, exercise caution to avoid plumbing and electrical components in the wall.

A tubing/pipe cutter is shown in Figure 2-18. To use it, place the cutting blade and opposing rollers over the circumference of the pipe, then screw the adjustment dial until the cutting blade is snug, and then next rotate the tool around the tubing, conduit, or thin-walled pipe to be cut. Score a light indention track around the pipe first, and then turn the adjustment screw again and rotate the tool again; repeat the process until the tubing is cut all the way through. When the pipe wall is thick, you can still use the tubing cutter to score a guide mark around the pipe for later cutting with a hacksaw. This helps keep your hacksaw on the mark for a straight cut. This tool will be helpful when cutting mounting pipe or braces to length or for cutting metal conduit to needed lengths. If your project includes the use of PVC conduit or pipe, your best bet is to cut it with a fine-toothed hacksaw.

Figure 2-19 shows a small hacksaw handle. Use this instead of a bigger standard hacksaw to cut things such as the excess from small screws after fastening them or for cutting thin pieces of metal. It is also a good way to get some more "mileage" from broken hacksaw blades.

From time to time, you might need to find out the exact size of a drill bit or bolt. A gauge card is shown in Figure 2-20 with a drill bit inserted. This card helps when you absolutely must have the right size hole drilled. Drill no hole before you know. Gauge cards are available in English and metric and for gauging wire sizes as well.

Figure 2-18 Tubing cutter

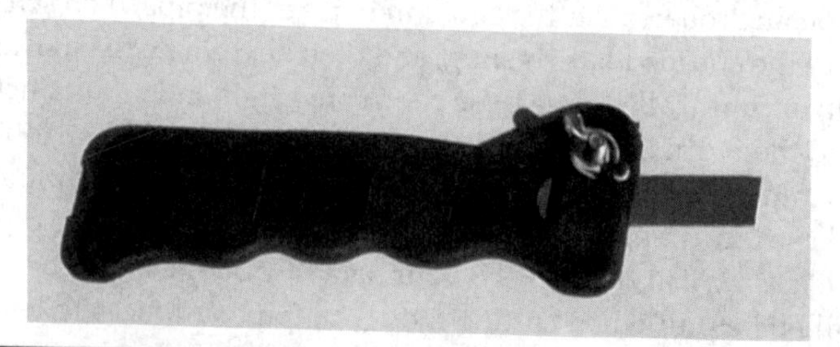

Figure 2-19 Small hacksaw blade handle

This will help you verify the drill size for screw pilot holes in wood or to drill the right size holes for machine screws and bolts. The drill bit's size stamp, if the bit shank has one, can be obscured while the bit is in the drill motor chuck.

For a small project with one or two cuts, using a handsaw (person-powered) will usually work out OK. However, it is a lot easier to use an electric-powered reciprocating saw such as the one illustrated in Figure 2-21, along with a blade assortment pack. Sometimes called a jigsaw or sabre saw, this versatile saw can be

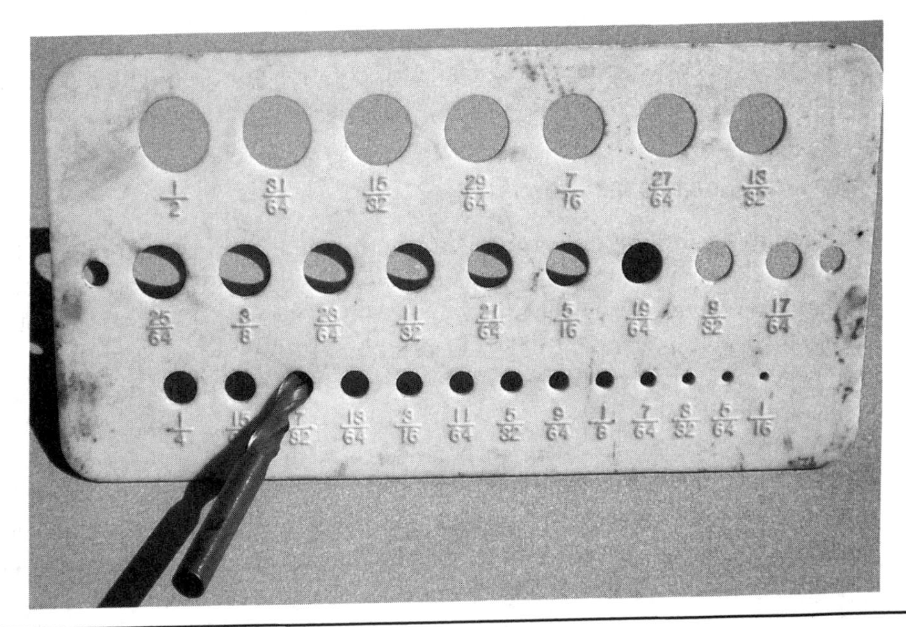

FIGURE 2-20 Drill and bolt gauge card

FIGURE 2-21 Electric-powered reciprocating saw and assorted blade package

fitted with a variety of cutting blade types for metal, wood, drywall, or plastic. The saw can be operated freehand to make freestyle cuts or used with a guide bar for long, straight cuts.

Measures, Squares, and Levels

To keep things on the up and up—I mean measured, square, plumb, and level—and to make sure everything is in its right place when you are done, you will want to have some measuring tools in your tool box for your FTA and any other DIY projects. If you have ever tried to level a large picture frame in an unlevel apartment or house, you probably already appreciate the importance of being able to measure things accurately.

Figure 2-22 shows a small "torpedo" level with bubbles for vertical, 45 degree, and horizontal alignments. It also shows three measuring tape rulers, with one in the upper-right corner that includes a bubble level. A traditional carpenter's folding rule is shown at the top, and a combination square with an analog level gauge is shown in the middle of the figure. Your FTA project will need a few of these tools

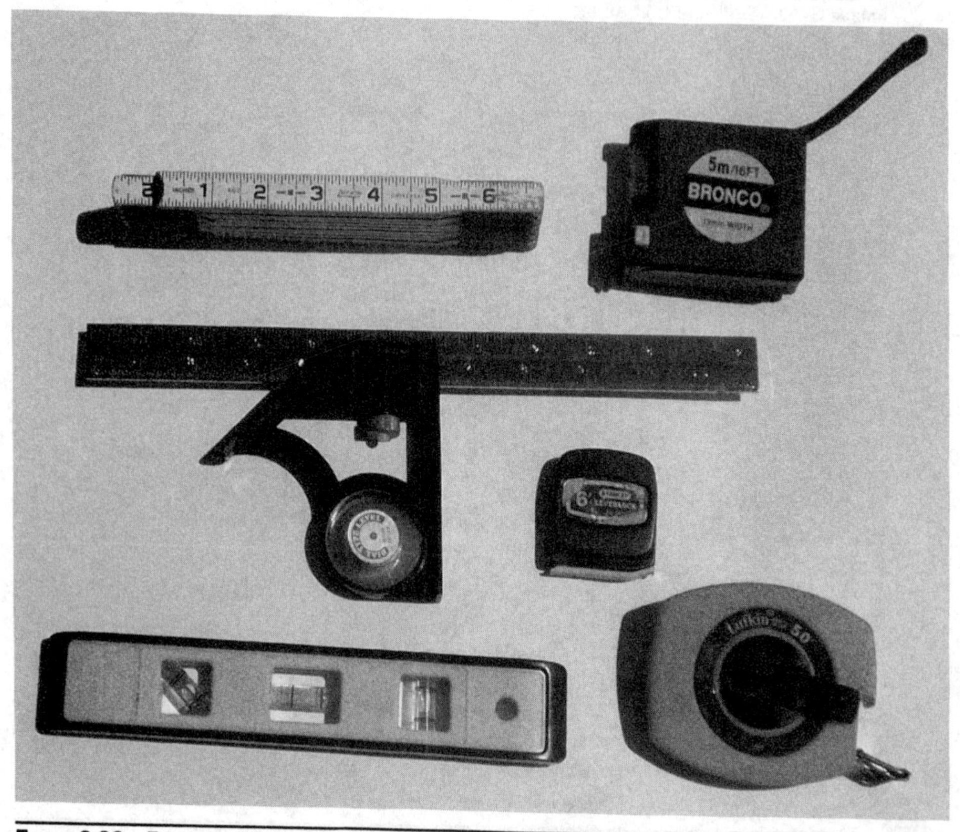

FIGURE 2-22 Examples of tape measures, combination square, and levels

Terms of Alignment

Just in case you did not already know, the term *level* means aligned perfectly with the horizon. As a reference, the undisturbed surface of water is thought to be level. *Plumb* means aligned with a vertical exactly perpendicular to the horizon at 90 degrees. A single drop of water falling from a stationary perch will in a draft-free space fall to the floor on a perfect vertical, or plumb, line. *Square* means that corners or line intersections are aligned at exactly 90 degrees in any observed plane.

for leveling up and locating drill holes. It is important that you keep antenna masts perfectly vertical, and these measuring tools can also prove handy when aiming the satellite dish up to a given satellite.

Specialized Tools

Unlike ordinary hand tools that perform well for multiple tasks and in a wide variety of project types, specialized tools are designed to assist you with a specific job or task and do it very well. Specialized tools tend to be used for various crafts and trades, such as carpenters, plumbers, or computer repair technicians. Some of the tools helpful with our projects for DIY installation of FTA and TV will be common in the tool pouches of satellite TV installers, cable TV installers, electricians, phone installers, and computer or network technicians. This does not mean, however, that you must have all the tools of these trades to undertake your project. But having a few of them in your tool box can be helpful and can keep your project on track to a speedy completion.

The tool shown in Figure 2-23 is called a crimper, but it can be used to do much more than crimp lugs and connections on wire ends, which is what the scissor ends do. The lug sizes with which it works are etched into the frame of the tool. Surrounding the lower portion near the fulcrum joint are five threaded holes for cutting small machine screws, which include the popular sizes for 8–32 and 10–32 (10 gauge–32 threads per inch) machine screws.

To use the cutter, thread the machine screw into the tool until the correct length is above the tool, and then squeeze the handle until the screw is cut. Unthreading the screw from the cutter repairs the bottom threads on the new shorter length screw. Just above the handle grips are a cutting blade for cutting wire and a gauged wire stripper for removing wire insulation. This tool is often used by electricians and auto repair mechanics. This versatile tool can be useful on some portions of your FTA or TV installation project, such as while you're running custom-made speaker wires for a five-point sound system.

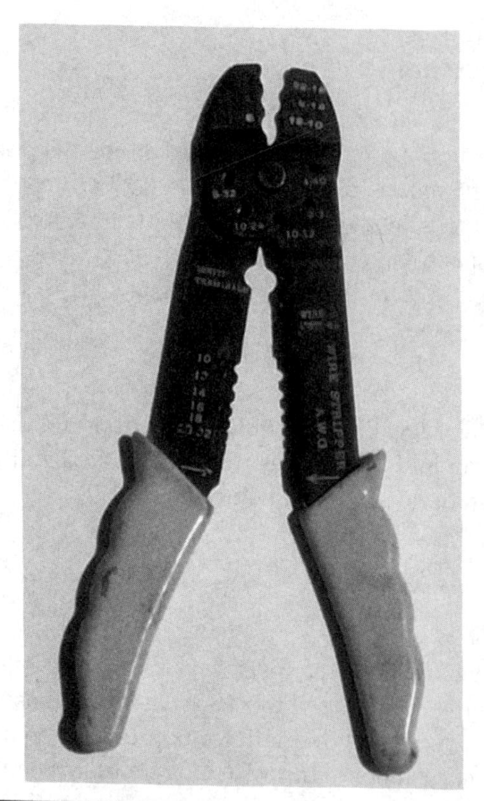

FIGURE 2-23 Crimper tool

A soldering gun is shown in Figure 2-24, along with a small spool of rosin core solder. This tool is useful for making permanent wire joints by soldering the copper conductors to lugs or to connections on equipment or wires to power jacks or plugs. Using a fine-gauge solder seems to work best for soldering on wiring connection projects. It melts fast, and you can control the amount of solder easier. Solder and the soldering gun can be used to make repairs or make new secure connections for such things as remote power systems or in low-voltage wiring in RVs or automobiles.

CAUTION *The soldering tip gets very hot to melt the solder. Never touch it. The piece you're soldering also gets very hot: allow the work piece time to cool off before you work with it again. Use a heat sink to protect wire insulation and sensitive parts while heating the joints with the soldering gun.*

Figure 2-25 shows two types of electrical testers. The plug-in version at left is intended to check for electrical power at an outlet and to verify whether the circuit is wired properly and check to see if it has a serviceable ground. The key-chart on the top of the tester is for interpreting the light sequence when it is plugged in. The tester at right is a neon light that will light if the test ends are connected to a wire

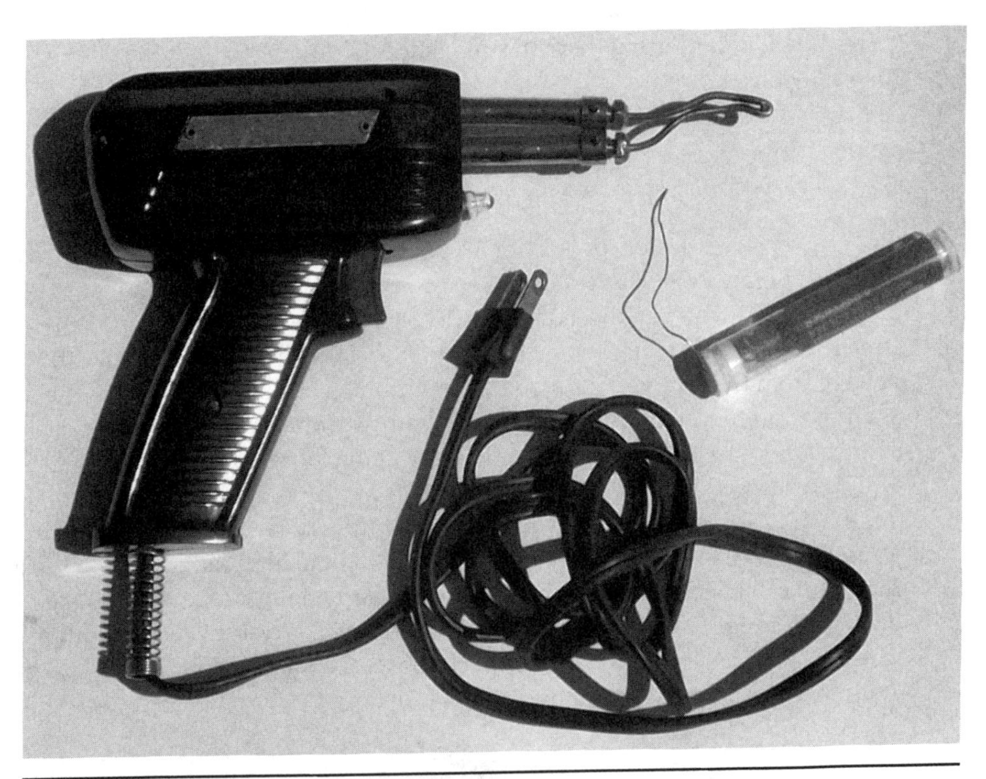

Figure 2-24 Soldering gun and solder coil

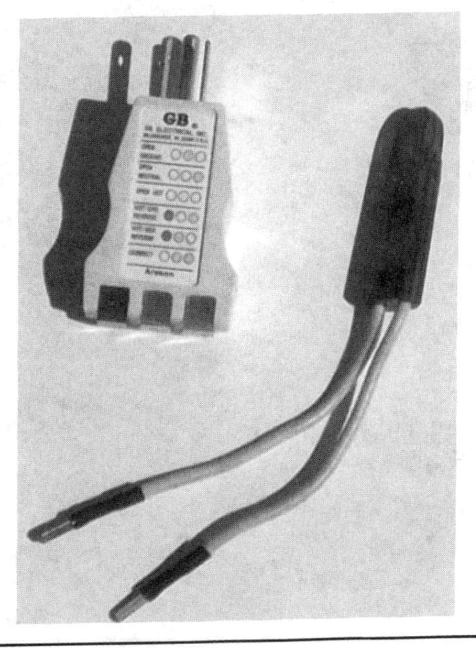

Figure 2-25 Circuit testers 120 volt A.C.

(or terminal) and return wire with sufficient A.C. voltage to light the neon, usually 90 to 120 volts A.C.

CAUTION *Never touch live (energized wires) with any part of your body. Always de-energize circuits at the breaker or fuse before working on them.*

A low-voltage tester for A.C. or D.C. voltages in the range of 5 to 50 volts is shown in Figure 2-26. This tester can be used on the RG-6 cable leading out to your dish to verify whether the nominal 12-volt current is present to run the signal strength meter as you use the meter to aim the satellite dish properly. Without that voltage from the receiver, the signal's strength meter will not function. If that voltage is not present, you might have a continuity problem in the cable or one of the connection joints might have failed or have come loose. The signal strength meter will be discussed in more detail in Chapter 7.

A handy size of a Volt-Ohm-Ammeter, or multimeter, is depicted in Figure 2-27. By moving the selector switch in the middle and matching the scale with the meter face, you can check for circuit continuity, measure an in-range A.C. voltage, measure an in-range D.C. voltage, measure milliamps, or test for the amount of resistance in a wire or circuit. If you plan to do a lot of DIY electronic projects, a multimeter is a must-have item.

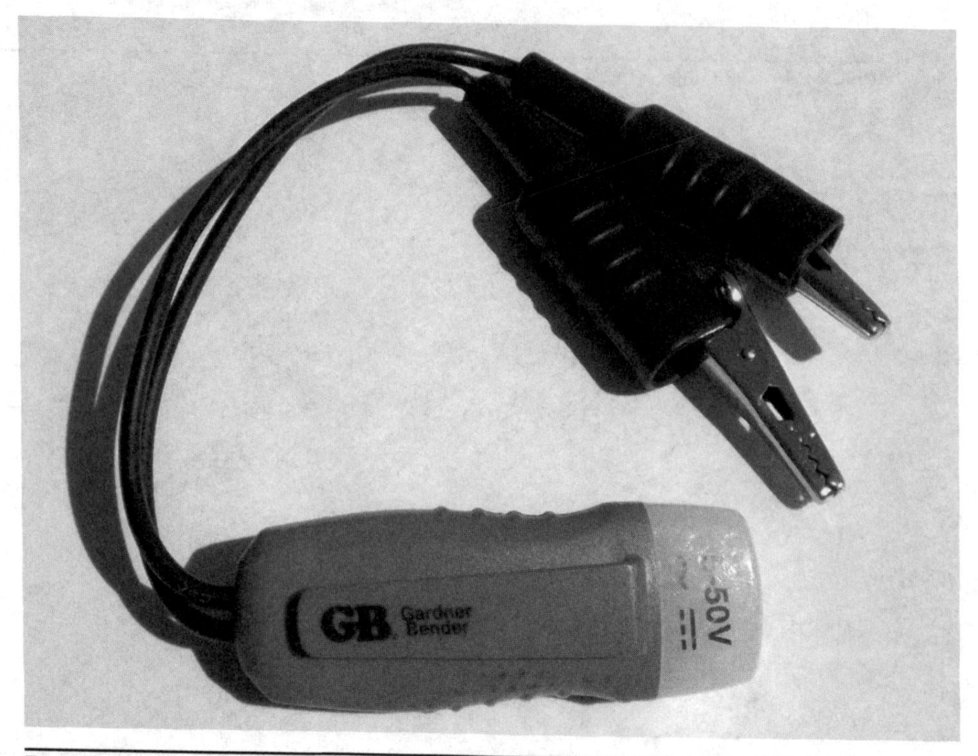

FIGURE 2-26 Low voltage A.C. or D.C. test light

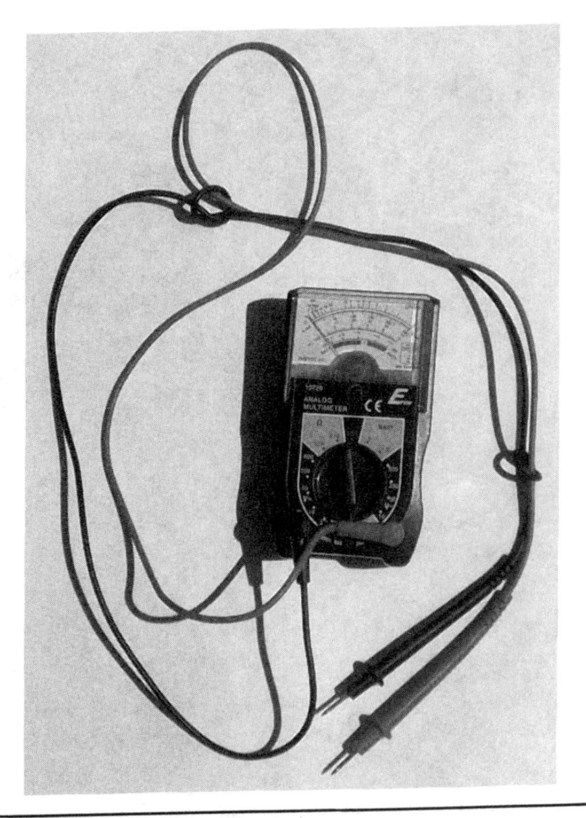

FIGURE 2-27 Volt-Ohm-Ammeter (a.k.a multimeter)

TIP *Multimeters run on batteries, so have a spare set ready. Rechargeable batteries are preferred, particularly if you store the meter for long periods, because they are less likely to corrode and ruin the meter.*

Not everyone will need to consider buying or finding the kit shown in Figure 2-28. If your DIY hobby requires that you do computer networking or telephone work along with FTA projects, you might want to own this networking tool kit. It includes testers for dial tone, network cable continuity testers, and stripping and crimping tools for CAT 5 cable ends.

The one component shown with the kit that you should use if you install a TV card in a desktop computer is the anti-static electricity shorting strap on the right in the figure. Before you begin working inside a computer case, disconnect the computer from all power sources. Connect the ground clip of the shorting strap to the computer case, and place the band on your wrist. Electronic devices are very sensitive to static electricity, and the shorting strap helps prevent damage from static sparks.

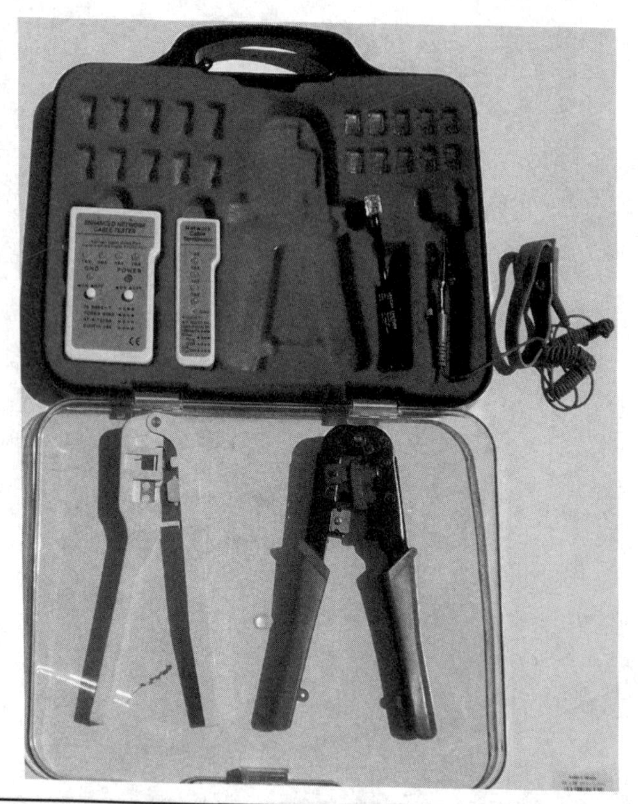

FIGURE 2-28 CAT 5 cable kit, phone connection crimping tool, and anti-static electricity shorting strap

Figure 2-29 shows important tools for any extensive RG-6 cable project. You can buy expensive cut-to-length pieces of RG-6 satellite TV cable with the cable terminations already completed, or you can buy rolls of bulk cable and the F-connectors separately and cut each piece to the length you need. The scope of the project that you are undertaking will help you determine whether the tools are worth buying. The more cable runs you do, the more likely you will benefit from having a coaxial cable crimping tool kit to install the F-connectors. The four-step crimping process for proper crimping requires the center conductor of the RG-6 coaxial cable (coax for short) to protrude a bit more than ⅛ inch beyond the connector. This process is covered in more detail in Chapter 5.

Miscellaneous Equipment

Figure 2-30 shows a pop rivet tool with a few pop rivets. Pop rivets are used by drilling a hole sized to match the rivet size and inserting the rivet in the tool, pressing the rivet through the hole in the material (such as fiberglass siding on an RV

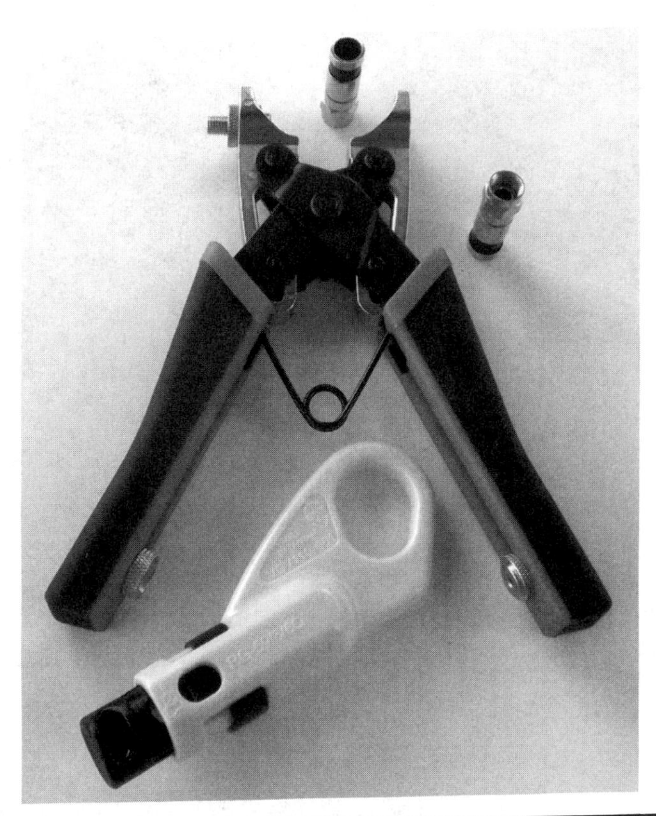

FIGURE 2-29 Tool kit for RG-6 cable: stripper (bottom), cable cutter (middle), and compression crimpers (top and upper-right)

and the thing you are mounting, such as a mounting plate for an external outlet for a RG-6 cable connection), and squeezing the handles of the tool together to clamp the rivet. That squeezing action pulls the rivet's pin out, leaving the back end of the rivet distorted and compressed against the siding material and the front of the rivet firmly pressed against the mounting plate, for a sandwich effect that holds the plate firmly in place without a high risk of cracking the fiberglass siding material.

To do work outside with power tools, you might need an extension cord, such as the one shown in Figure 2-31. Always plug in cords from the load (tool) back to the source (outlet) and reverse the process when you're done: unplug the cord from the outlet first, and then unplug the load (equipment) from the cord.

Always use a ground-fault interrupter circuit (GFIC) outlet when working on basement floors, in the garage, or outside. If you do not have a GFIC outlet, you can use an in-line outlet such as the one depicted in Figure 2-32. The GFIC is intended and designed to work and protect you from shock, even if there is no bonding ground wire in the house circuit.

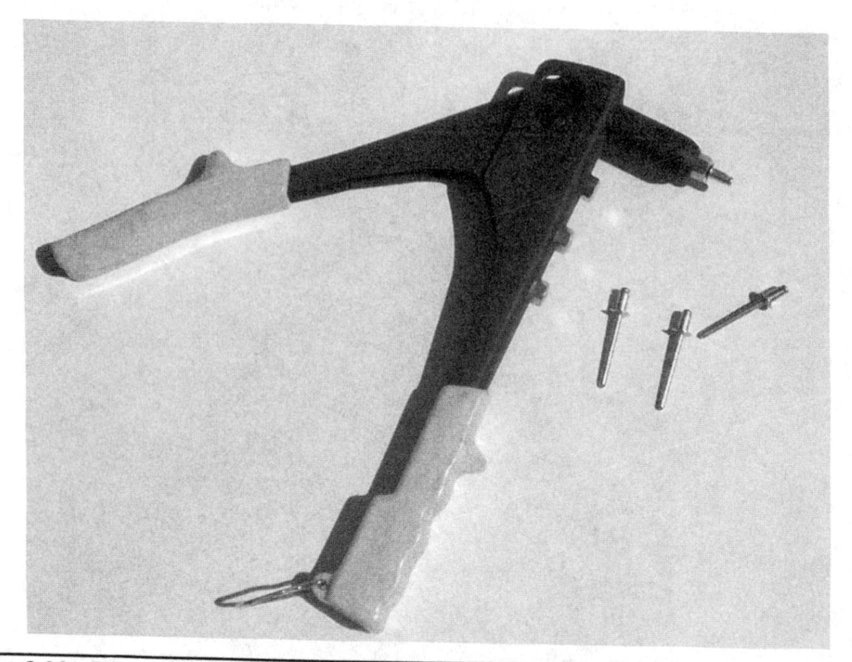

Figure 2-30 Rivet gun and three sample rivets

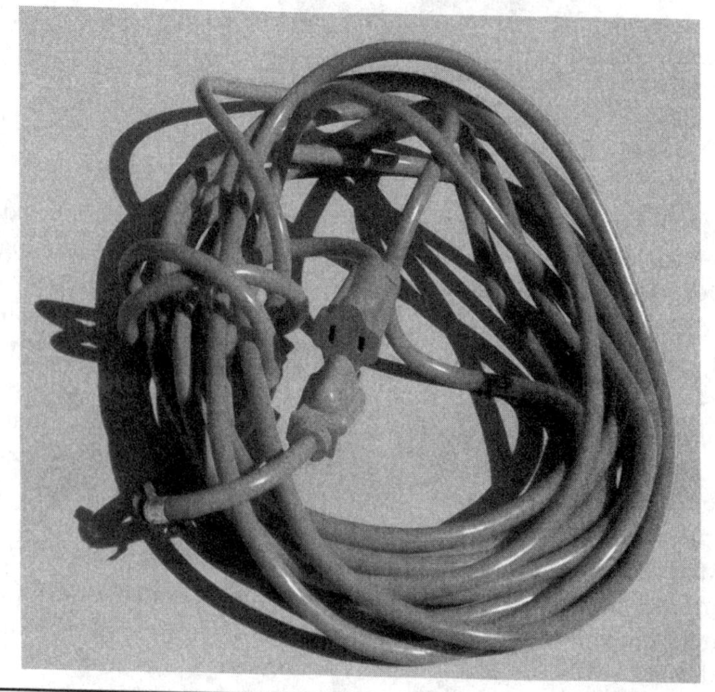

Figure 2-31 Extension cord

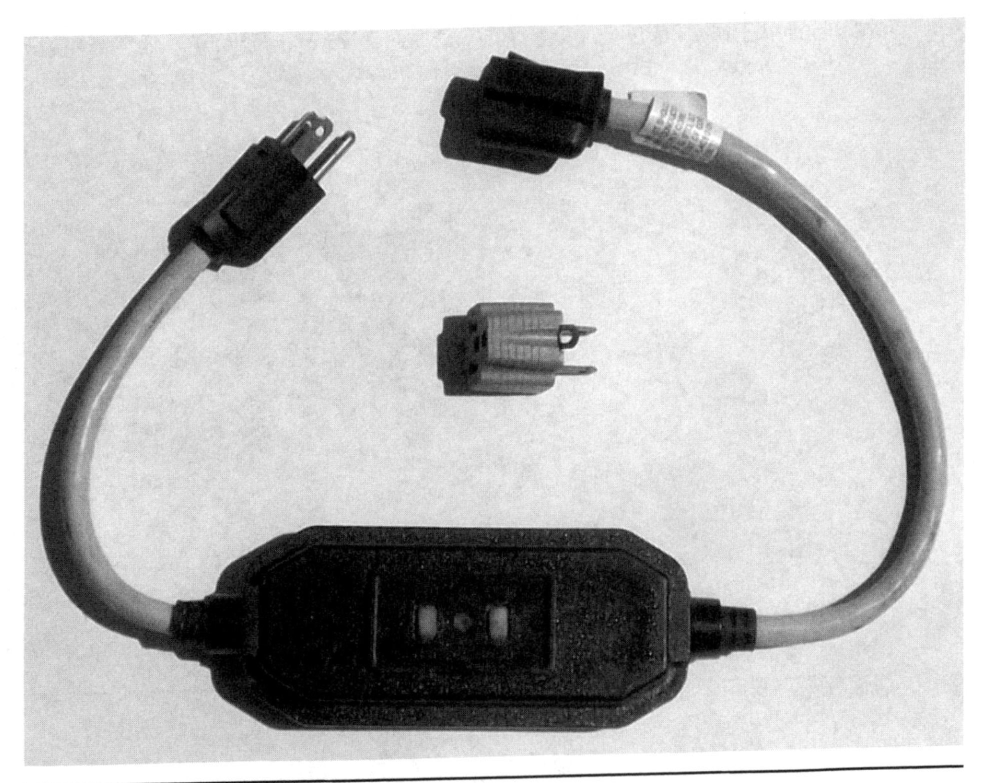

FIGURE 2-32 In-line GFIC outlet

NOTE *Some GFIC protection devices can fail, so always use the test button to determine serviceability before use. Replace it immediately if the test fails.*

For working in dark places, such as under a porch, you might use a trouble light, like the one shown in Figure 2-33, to light your way. Trouble lights for safety should also be plugged into a GFIC outlet or in-line outlet.

Conclusion

You might not need every tool covered here to perform a successful installation of an FTA satellite system. For some readers, the basic tools, specialized tools, and miscellaneous equipment listed here are not nearly enough tools for stocking the DIY project tool box.

Take a few minutes to plan out your project and think of the tools your unique project will require to complete. Collect them in one place and get ready to get started on the FTA installation.

NOTE *Remember that safety should be first in the tool box and the first "tool" taken out. Keep that "safety cap" on while working on this and every other DIY project.*

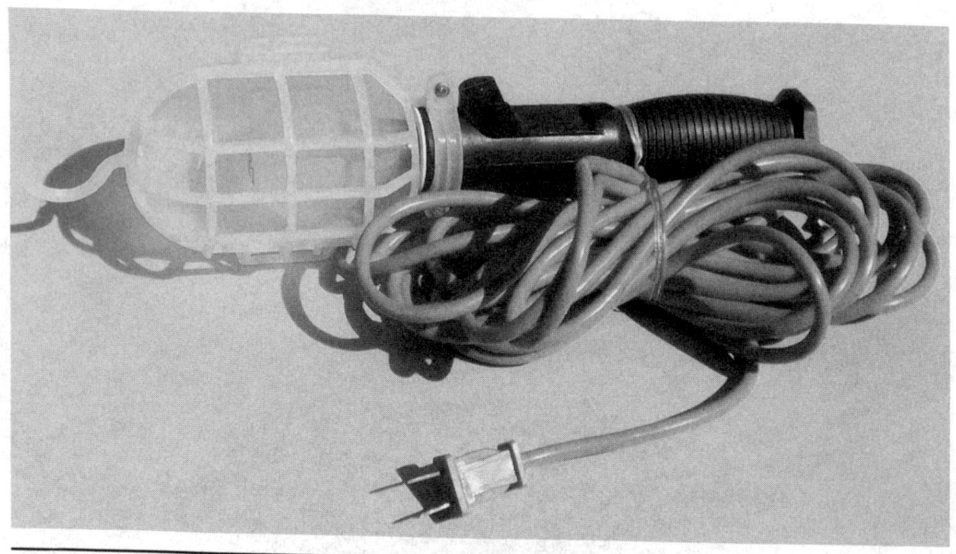

FIGURE 2-33 Trouble light

CHAPTER 3

What's Out There? Available Free Channels and Satellites

Getting television and video programming along with music and talk programming for free—meaning *no monthly fees ever*—is the entire single-minded focus of this book. After you bear the modest cost of the FTA receiver and take the time to set it up properly, the only cost of FTA operation is a very modest amount of electrical energy necessary to run the TV set and satellite receiver system.

Electrical power consumption will vary with the size of the TV and additional equipment. On the low end, a 30-watt FTA receiver and a 70-watt 22-inch flat screen TV, if left on 24/7 for a year with an electric utility power cost of 15 cents per kilowatt hour, would cost less than $150 per year, or about $12.50 per month. With the same TV connected to a paid-for cable or satellite box, the cost will be similar for electricity, but of course you'll pay more for the cable or satellite service.

FTA's minimal setup costs and the fact that the service is free is the very biggest upside of FTA and OTA television. Somehow, viewers have been convinced that spending $30 to $100 per month, and more, is a logical step to increase viewing options. Most TV viewing occurs in the evening hours, typically in the range of 5 to 6 hours a day per adult viewer. A program package of 200 channels from cable or pay-for providers will likely include stations in the cost that will never be watched by anyone in the household. Another benefit of FTA and OTA TV is that never watched programs and always watched programs are the same price month after month: free.

There is a downside to FTA programming and station viewing options, and that is simply stated by the well-worn cliché: Here today, gone tomorrow. There is no guarantee that any particular program channel that can be viewed today will be available on FTA forever. Stations and channels come and go. New ones are added and old ones go away. There is little or no warning to either event. You'll find new ones when you do a new blind scan, or you'll read or hear about a new channel from a fellow FTA viewer. The ones that go away will come up blank on your screen when you look for them. But this is not a huge problem, and it's not like they all disappear at once and that nothing will ever replace them.

Two factors influence what satellites anyone in the United States can use to receive FTA programming: the quality of the equipment, particularly the dish and setup, and the geography of your receiving locations both determine availability. Because of the Earth's curvature, East Coast viewers can receive signals from satellites that West Coast viewers cannot receive, and vice versa. Signals will be stronger in Southern states than in Northern states, significantly enough to affect reception quality, and a larger dish or more efficient low noise block (LNB) might be necessary in northern latitudes.

With all these things in mind, the rest of this chapter will present information that represents the kinds of programming that is available on FTA from many of the FTA channels that can be received from most areas across the United States. Readers living outside of the range of the longitudes of 66°W to 140°W will have a different mix of target satellites and program viewing options and might require a different LNB with a lower input frequency.

The information presented here is based on scans at specific points in time and requires this disclaimer:

> Your experience and what you can receive will be based on what is available when you set up your system, as channels come and go on satellite transponders. Your experience can also be different based on your dish size, low noise block feed (LNBF) frequency range, the quality of the equipment, and the quality of the installation. What I know at this writing from my own experience is that 70–100 channels of English language programs and hundreds of foreign language channels are available on FTA. The variance is based on what you count. For example, if a South Carolina PBS station has multiple channels and major network news feeds are on more than one channel, but they come and go based on the network's need, how do you best count them? Some web sites list the programming and channels.

A big part of the fun of FTA is the quest to discover what is out there—not only regarding channels available but regarding what is being presented on those channels. The neat part is you can get a real taste of things foreign without ever leaving home. It's like being on a virtual voyage across the globe. All it takes is a few presses on the remote to go from Bolivia to Russia and see what they are seeing on their news, and hear and see the universal language of music being performed live for TV from all corners of the globe.

The English programming originating in other countries will present a world viewpoint you might not have appreciated before listening to them for a while. Even if you disagree, this information is still informative and educational, sometimes downright thought-provoking. The exposure possible with FTA hitherto has been only within the exclusive reach of diplomats and those with large travel budgets or the handful of longtime FTA hobbyists and viewers across the country. With FTA, it is not quite the same as being there, but it is fun and educational anyway.

Satellite Channel Reception in North and South America

Launching a satellite into orbit and maintaining it throughout its lifespan is an expensive proposition that can cost in the neighborhood of $150 million and more, with a third of that cost associated with the rocket that is fully expended on launch. Figure 3-1 shows a specialized communications satellite launched from the Cape Canaveral AFB in May 11, 2011.

The launches are amazing, but what is more amazing is what these car-sized bundles of silicon do once placed in a geostationary orbit. It is even more amazing that you and I can benefit from past launches and future ones for just a few hundred dollars spent on FTA receiving equipment. The best news of all is that more than 33 satellites carry programming of potential interest.

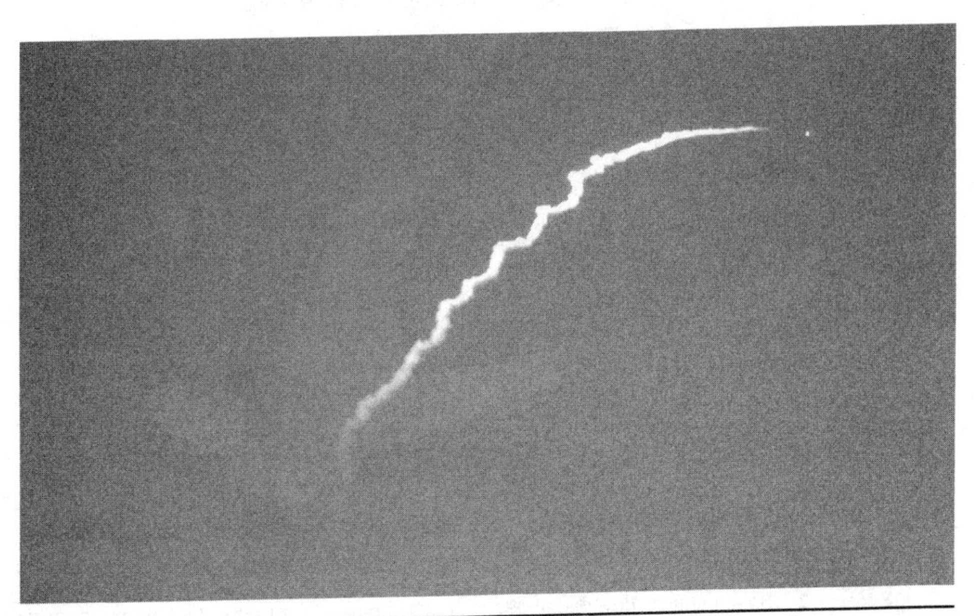

FIGURE 3-1 Rocket launch from the Cape Canaveral AFB May 2011

NOTE *Ignoring this opportunity and leaving FTA for others reminds me of the cliché, "tripping over dollars to pick up (save) pennies." Spending $300 on a system to view programming from only one satellite leverages the expense of each dollar spent on each commercial launch by about 50,000 times.*

It takes a great deal of additional energy, as demonstrated by the fireball size in Figures 3-2 and 3-3, both taken quickly after launch, to get space shuttles into tem-

FIGURE 3-2 Space shuttle rocket engine fireball seconds after launch

FIGURE 3-3 Space shuttle downrange east from Kennedy Space Center less than a minute after launch

porary orbit, making rockets the preferred method for satellite payloads. With the shuttle vehicle program on hold, rockets will be the only satellite delivery method for at least the next decade.

When you watch a rocket launch with a satellite payload or a shuttle launch, you can feel the energy in the air and sense the light from the engines glare on your skin. The technology that gets the satellites and other payloads in the air is no less amazing than what they do once in orbit or on mission.

The following numbers of possible channels for these satellites are based on scans in May and June 2011 and are intended to be representative and not an exact number of channels that will always be there.

Satellite by Name	Number of Channels	Satellite Position
Telstar 14	27	63.0°W
Star One C1	8	65.0°W
Star One C2	49	70.0°W
AMC 6	17	72.0°W
Simon Bolivar	22	78.0°W
Nimiq 4	2	82.0°W
AMC 9	2	83.0°W
Brasilsat B4	46	84.0°W
AMC 3	85	87.0°W
Galaxy 28	17	89.0°W
Galaxy 17	64	91.0°W
Galaxy 3C	15	95.0°W
Galaxy 19	289	97.0°W
Galaxy 16	31	99.2°W
SES 1	59	101.0°W
AMC 1	37	103.0°W
AMC 15	8	105.0°W
Galaxy 18	56	105.0°W
Satmex 6	20	113.0°W
Satmex 5	67	116.8°W
Anik F3	8	118.8°W
EchoStar 9 and Galaxy 23	42	121.0°W
Galaxy 18	9	123.0°W
Galaxy 14	52	125.0°W
AMC 21	9	125.0°W
Galaxy 13	2	127.0°W
AMC 11	3	131.0°W
Galaxy 12	2	133.0°W
AMC 10	8	135.0°W
AMC 7	2	137.0°W
AMC 8	10	139.0°W

FTA Highlights by Programming Type

If you had to find one word to summarize what is out there by way of programming on FTA satellite broadcasts, it would be variety. It is fair to say that there is a little bit of everything; the familiar and the unfamiliar in video and music. One of the first channels I found extremely interesting after setting up my system was Greek performers playing instrumental-only music. It was not something I expected to find on television. The thought I had was this is too good for TV, wish I could be there. Here is a short summary of what you might find on FTA.

- **Foreign language** Foreign language broadcasts are too numerous to detail. Within foreign language broadcasts you will find audio-only feeds, music, and news in many languages. English language broadcasts originating in other countries such as Russia, China, and the Netherlands are also interesting to watch.
- **Religious** The religious programming on various satellites runs the gambit from Life Talk Radio, to Gospel music, to God's Learning Channel.
- **News and political** Channels of political interest include English language channels from Saudi Arabia, Kuwait, Iran, European countries, and South American countries. Network real-time live news feeds of all sorts can give you the news as it happens, not just at 6 P.M.
- **Educational** Children's programming, business TV, history, biography, college stations, and a bonanza of PBS and university programming of all sorts (from Montana, Oklahoma, Washington, Alaska, Florida) round out a generous offering of educational material on FTA.
- **Networks** Major network affiliates are often rebroadcasted on satellite for the benefit of inputs for cable providers and can include familiar networks such as FOX, CNBC, ABC, NBC, and CBS stations. There are also many networks available that you might not recognize as well.
- **Movies and TV shows** Movies and programs from old to new, and from English to nearly any language can be found. Try watching a movie you have already seen that's been dubbed in Spanish to test your Spanish skills.

FTA Highlights by Network Logo or Call Sign

The list of logos and call signs presented here is not at all exhaustive or complete as far as what can be found on FTA. This listing is meant to be a representative sample to demonstrate what you will find familiar on FTA right along with strange and unusual. It is too fun pressing a button a few times and being surprised or impressed after a blind scan on a new satellite. Fortunately for my past, present, and future readers, I find it just a little less fun than writing. It is always amazing to find new things on FTA TV and often entertaining enough to watch for an hour or two at a time. See if you can find a familiar entity in the alphabetical listing that follows. All the logos and trademarks are owned by their respective registered titleholders.

A	ABC, Al Jazeera, Ariana, Azcar
B	Bio, BYU, Bloomberg TV
C	Church Channel, Cornerstone, Create, CW, CBS, CYC
D	Daystar
E	EBRU, Emmanuel TV
F	Fox 64, Fox 19
G	GLC, GBN, GM
H	Hope, House of Yahweh, History
I	ION
J	JCTV
K	KATV, KBTZ, KCBU, KFDF, KFTL, KKYK, KLMN, KBPI, KQDK, KQUP, KTVC, KWWF
L	LLBN
M	Macy's, MY42, MY 31
N	NYN, NTA, 360 North, NBC, NASA
O	ONN
P	Patient Channel, PBS, Peace, Pentagon, Press
Q	QVC, Qubo
R	Research, RT, RTN, ReelzChannel
S	Safe TV
T	TBN, Telemundo, TUFF TV, TV Guide
U	UWTV
V	Vme
W	The Weather Network Channel

Discovery

Once you have your FTA receiver and dish antennas set up, you can set about discovering programming and channels on these birds that are of interest to you and your viewers. The offerings will change without notice, which is part of the fun. The receivers can be set to rescan a given satellite while you are away doing something else. With 30 or so satellites of possible interest, scanning one each day after the initial receiver setup will keep your program/channel lists relatively up to date.

The information here hopefully will entice you to explore the value of having your own FTA system and see the value proposition presented by FTA to supplement OTA. In addition, perhaps you'll find enough programming offered to drop the high priced pay-for-services delivery models present in cable and satellite services. It is also possible to co-exist an FTA receiver with the pay-for-services offerings simply by plugging in a few cables to your set from the FTA receiver. In other words, you can run them both for a while to try before you "un-buy" the paid services.

Advertising and Pay-for-Service Products

Nearly all products and services we buy today include some portion of advertising costs. It would be nice if getting FTA and OTA programming from those purchases would be the norm. I am particularly turned off as a consumer by the idea of paying a cable or satellite service provider so that I can be shown an excessive number of advertisements per program on whatever channel I watch. For the products or services that I buy that advertise only on premium subscription models, that portion of the product or service's cost is considered a total waste. For me and my viewers, FTA will always be a part of our programming lineup. It's hard to find a better deal anywhere. Eventually, advertisers will find FTA channels much like they have found the Internet. Regrettably, advertising on the Internet may not have provided any noticeable improvements in the programs or content offered. The content on FTA already presents a reasonable value proposition: it is my hope that when the advertising dollars follow the viewers as they embrace FTA, that FTA will improve and offer even more variety and value to viewers in the future.

Conclusion

To set the record straight, I am not against cable companies and the pay-for satellite programming companies. They provide a service that many consumers believe to be of value. I am unapologetically opposed to the high prices charged and the lack of consumer choice. I know from my work in the computer security field that all the technology components are in place for the pay-for providers (cable or satellite) to give each household only what they want in programming and that to bill accordingly is technically possible. These initiatives are called "al a carte" packages or "channel choice," but legislation on choice initiatives tends to go nowhere because of industry pressure to maintain the status quo.

Another stumbling block is programming contracts that impact what is offered on pay-for services systems. As consumers of television programming, we do have a choice and that is to switch from the expensive plans and use FTA to supplement OTA. If enough households do this, a giant elephant will be in every living room and the companies will eventually figure out that lowering costs and prices while improving consumer choice is necessary to compete in this new global information economy.

As for the issue of what to watch on TV. I believe that market forces in this global economy can finally benefit the average household in a big way if enough people switch to FTA. As I write this I am listening to CCTV 9 News from China in English from the satellite at 91.0 degrees west. The programming is informative, interesting, educational, and entertaining, and I do not have to listen to what seems like senseless political rants that appear so frequently on American networks. When I grow weary of this channel I will press a few buttons and find out what is going on south of the border or watch some Latin drama show.

The world is my showcase with FTA; it's the best (at least in the top ten best) $300 I have ever spent. The world's free programming can be at your fingertips, too, with your own FTA system.

CHAPTER 4

Antenna Basics

In this chapter you will learn about how antennas catch the signals from a broadcast source and send them to your receiver's tuner/decoder, which processes them into the programming you want to watch. The text will help you decide which antennas are best suited for your needs and will provide information about how to assemble a satellite dish antenna for installation.

How Antennas Work

An antenna's purpose is to catch enough of the transmitted electromagnetic waves to provide an electrical signal level sufficient for the receiver to process the signal into useful audio and video. In the sections that follow, the text will briefly discuss how this works in simple terms.

Signals Have Sources

Radio waves (*signals*) are electromagnetic fields emitted from a natural source, such as a lightning strike; from an incidental manmade source, such as the arcing brushes of a running electric motor; or from a purpose-filled source, such as a television station's transmitter tower. The TV tower's transmissions are unique in that its signal is designed to carry useful information embedded in the radio wave.

Any time a magnet or magnetic field is passed over a wire, electric current flows through the wire. An *electric current* is simply electrons moving from the outer shell of one atom in the wire to another atom in the wire. Conversely, any time an electric current is passed through a wire, a magnetic field is present. Radio waves (TV signals) are made up of these fields of energy that combine an electric force (moving of electrons) and its companion magnetic force fields.

The antenna captures signals from the manmade sources and sends these signals to one or more of our FTA, OTA, or broadcast radio station receivers. Two physical variables of electromagnetic (radio) waves impact antenna design:

- The strength of the emitted wave, usually measured in watts or milliwatts
- The signal's frequency, measured in Hz, KHz, MHz, or GHz

Signal Strength or Power

You have probably heard radio stations advertise that their antenna tower puts out 100,000 watts, which would mean that the station has a very far-reaching signal in comparison to one that has only 10,000 watts of radiated power. Station signals can be weak or strong from any point on the planet. The further the receiving antenna is from the transmitting antenna, the weaker the signal. The signal's strength weakens with distance, much like sound or light does. FTA satellites orbit the Earth in sync with the rotation of the Earth at an altitude close to 22,236 miles above the equator. Unless the dish antenna is at the equator, the signal has to travel more than the 22,236 miles from the transponder to reach the receiving dish—that's a long way to go to reach the ground-based antenna from a satellite transponder. The output power from satellite transponders is low compared to Earth-bound transmitters, on the order of only 20 to 120 watts.

Frequency

To receive your favorite FM station, for example, you might have to tune in 98.1 MHz, the station's frequency. Frequencies are either high or low, relative to each other along the entire electromagnetic spectrum. When a transmitting station is broadcasting with a high frequency signal, the wavelength is short. Stated another way, with higher frequency signals, the next oscillation or cycle of the wave follows closely behind the previous one. This closeness of the succeeding waves is particularly acute from satellite transponders. In the Ku band, for example, the frequencies are in the 10 to 12.5 GHz range, so the wavelength is very short.

Wavelength

Radio waves and light waves travel at the same speed, regardless of their frequency: they travel at the speed of light. Light travels nearly 300,000,000 meters (186,000 miles) per second. Wave length is the distance the first wave travels before the next one starts its journey away from the transmitting antenna. The wavelength for an FM station at 98.1 MHz on an auto's radio dial is 3.058 meters (10.030 feet). An AM station at 920 KHz on the dial's wavelength is 326.08 meters (1069.56 feet), a much longer wavelength than the FM station. The wavelength for a satellite transponder on the Ku band transmitting at 12 GHz is very small, at 0.025 meters (0.984

inches—less than 1 inch). The formula to calculate wavelength when (f) frequency is known is to divide the 300,000,000 meters per second by the frequency in Hz as the denominator, and the quotient result is the (wl) wavelength in meters (300,000,000 m./f Hz = wl m).

Antenna Size

The ideal long-wire antenna length is one full wavelength. It would be impractical, if not impossible, to have the perfect length antenna for every frequency of the stations or channels to which we want to listen or watch. Even if a half-wave dipole antenna were used for each frequency, it would take a lot of space to have the antennas tuned to each frequency.

Because the satellite transponders are transmitting high frequencies at such great distances and with low wattages (power) in comparison to an Earth station's transmitters, it would be impossible with today's technology for a receiver to receive enough of the satellite signal with a 1-inch antenna. Power levels for transmitting stations are limited by regulation and, in the case of satellites, are also limited by the physics of the available power supply. So multiband (multifrequency) antennas of any sort are a compromise to begin with, because it is impossible to be changing the length or size of the antenna constantly to match the receiver's dialed-in frequency settings. Earth-bound transmitters are emitting strong enough signals to cope with the design limitations of the receiving antennas. Receiving satellite signals of sufficient strength on Earth is accomplished by the dish's unique design and by its size. In regard to satellite dishes, the cliché "bigger is better" isn't the case: the truism is "big enough is necessary." Trying to receive FTA signals from distant satellites with an 18-inch dish simply does not work out particularly well.

Gain

To receive ground-based OTA TV stations that are far away, or signals from satellites in orbit, antennas are designed to be very effective and are rated by the amount of *gain* (directional efficiency) relative to a theoretical *best* antenna—usually a half-wave dipole antenna. Gains, and losses, of signal strength are expressed in decibels. (For example, some of the received energy is lost traveling down the cable to the receiver.) Gain is express in positive numbers, and losses are expressed in negative numbers. Decibels are measured as a ratio exponential on a logarithmic function: *10 × log(power out/power in).*

TIP *For readers who do not want to get wrapped up in the math, when you see a design characteristic expressed in dB, simply remember that it is not a linear function. For example, a 3 dB gain means that the signal is 2 times the reference signal's strength. Also keep in mind that measured or stated dB gains or losses, when used to describe an antenna's efficiency, are at a specific test frequency, and if the tuned-in frequency changes, so will the gain or loss ratio.*

The Antenna's Work

Transmitting antennas, including transponders on satellites, push electromagnetic waves out and away through space and into the Earth's atmosphere. When the waves strike the receiving antenna dish on Earth, they are reflected to the low noise block (LNB). The energy in the captured waves induces a detectible and measurable electric current in the antenna's LNB that then travels down the cable to the receiver. This induced electric current collected by any antenna is in nearly exact proportion to the transmitted signal in regard to frequency, and with the embedded data it carries through frequency modulations or amplitude modulations. Of course, the signal that is received is much weaker than the signal transmitted, but it has to be strong enough for the receiver to amplify and decode into a video and an audio signal. The transmitted signals for audio and video television programming are said to be digital, meaning that the signal encoding are 1's and 0's, like a computer's internal workings or the data travelling over the Internet. OTA TV signals are also in digital format now in the United States and Canada, which makes high definition television (HDTV) broadcasts possible and practical.

Antenna Variations

To say there are more differences in antenna designs than one could imagine might be an understatement. Earlier in this chapter I mentioned long-wire antennas and the variation in length necessary to tune one to a given frequency. In every case when an antenna design element or characteristic is changed, there are solid engineering principles being applied giving rise to the variation. In the sections that follow, the text will discuss a few of the principles that give rise to the necessary variations in antenna design.

OTA TV Antennas

It is interesting to note the changes that would be necessary in antenna size to match the wavelengths of the various OTA channel frequencies. As stated earlier, antenna design involves compromises in the design. Antenna designers and manufacturers work hard to make antennas that will perform well over a wide range of channel frequencies. The sections that follow demonstrate the different lengths that would be used for a half-wavelength antenna element at example channel frequencies involving broadcast television.

VHF Antenna Element Size for Half-Wavelength at Midchannel Frequency

Notice how dramatically the wavelengths change as the channels change. See the table on page 50.

Legal Issues

In the United States, we all have a legal right to receive paid-for programming and no-fee FTA satellite signals, regardless of what a deed covenant, homeowners' association, or landlord might have you believe. Even if you are not a dyed-in-the-wool Federalist, sometimes it is helpful to point to federal laws and regulations when they help ordinary citizens achieve their goals or exercise basic rights. Should you find yourself living in areas of intense zoning, planned communities, or housing developments with deed covenants or restrictions, or in a neighborhoods with out-of-control homeowners' associations, you should know that the Federal Communications Commission (FCC) permits use of satellite dish antennas of 1 meter (39.37 inches) or less in diameter in the lower 49 and of any size in Alaska. The quote that follows is directly from the FCC fact sheet on the topic (www.fcc.gov/mb/facts/otard.html):

> The rule (47 C.F.R. Section 1.4000) has been in effect since October 1996, and it prohibits restrictions that impair the installation, maintenance or use of antennas used to receive video programming. The rule applies to video antennas including direct-to-home satellite dishes that are less than one meter (39.37") in diameter (or of any size in Alaska), TV antennas, and wireless cable antennas. The rule prohibits most restrictions that: (1) unreasonably delay or prevent installation, maintenance or use; (2) unreasonably increase the cost of installation, maintenance or use; or (3) preclude reception of an acceptable quality signal.

This is not a *carte blanche* rule that allows a renter to destroy or modify rental properties or take over common areas, but it at least provides a beginning for negotiation with less-than-accommodating landlords and a regulation to show the homeowners' association boards if someone takes exception to a dish or other antenna installation.

A practical tip for dealing with no drilling situations is to use a special flattened cable designed to fit under a casement window, as shown in Figure 4-1. It is still necessary to provide a proper Earth/service ground connection for an antenna cable brought into a dwelling with this type of lead.

Figure 4-1 Flat cable for use under windows

Frequencies	Channel	Distance
054.000–060.00	Channel 02	8.6 feet (2.63 meters)
060.000–066.00	Channel 03	7.8 feet (2.38 meters)
066.000–072.00	Channel 04	7.1 feet (2.17 meters)
076.000–082.00	Channel 05	6.2 feet (1.89 meters)
082.000–088.00	Channel 06	5.7 feet (1.76 meters)
(Frequencies between 88 and 174 MHz are used by other services.)		
174.000–180.00	Channel 07	2.8 feet (0.85 meters)
180.000–186.00	Channel 08	2.7 feet (0.82 meters)
186.000–192.00	Channel 09	2.6 feet (0.79 meters)
192.000–198.00	Channel 10	2.5 feet (0.76 meters)
198.000–204.00	Channel 11	2.4 feet (0.74 meters)
204.000–210.00	Channel 12	2.3 feet (0.72 meters)
210.000–216.00	Channel 13	2.3 feet (0.70 meters)

UHF Antenna Element Size for Half-Wavelength at Midchannel Frequency

UHF is in the range of 470.000 to 512.00 MHz as the frequency band for broadcast TV channels 14–20, at 6 MHz increments for each channel. The element width for channel 16 at midrange would be 473 MHz, and the element width would be 12.5 inches (0.32 meter).

Now let's consider broadcast TV channels 21–69 in the assigned frequency range of 512.000 to 806.00 MHz. Picking just one channel, 68, near the end of this range would yield a midrange frequency of 797 MHz and a half-wavelength element of 7.4 inches (0.18 meter).

FM/AM Antenna

From these few examples, it is easy to see the great challenge involved in designing an antenna to work well over the vast differences in bandwidth. Improving radio broadcast reception over greater distances requires a receiver with a port for accepting an external antenna, and an outdoor antenna. If you live where outdoor antennas are not possible, you can consider one of the amplified indoor antennas instead.

Consider antennas that are easy to mount and install. A vertical whip antenna, amplified or unamplified, of at least 50 inches tall with a mounting bracket is a good starting point. Amplified version antennas require a power source. To make this type of antenna most effective and improve performance, mount it at least 25 feet above ground. Use a shielded down-lead and install a properly grounded lighting arrestor. Grounding wire is usually bare wire and should be a minimum of 6-gauge wire.

NOTE *See Appendix A for some sources for broadcast radio antennas. For more current information and source links, you can also visit the author's website at www.allaboutfta.com.*

Combined UHF/VHF

Several combined antennas on the market will perform reasonably well for receiving channels 2–69 for receivers located less than 30 miles away from the transmitting tower. Omni-directional antennas, double-bay bow-tie antennas, quad-bay bow-tie antennas, and even old-fashioned rabbit ear antennas can work if you are close enough to the OTA station towers. It is also possible to make your own quad-bay bow-tie antenna from a piece of wood, old metal coat hangers, some wire, and a 75–300 ohm matching transformer (a *balun*).

To receive OTA stations from distances of 30–65 miles will require multi-element antennas with high counts of both UHF and VHF elements included on the boom (horizontal center post). Some of the elements measured end to end will reach 110 inches. The boom will reach 130 inches and longer. Usually these larger antennas are mast mounted and are accompanied by a rotor motor on the mounting pipe so the antenna can be pointed at source stations by using a remote-control in the TV room. If you are far from the stations, you might need to use a large, highly engineered antenna which is expensive to set up.

Amplified Antennas

For OTA reception of local and near distant television channels, a wide-band amplifier can be incorporated into the antenna. Antennas attached on the top of RVs and travel trailers are amplified antennas, as shown in Figure 4-2. The RG6 cable leading up to this antenna carries 12 volts from the switch box, which is connected to the RV's 12-volt DC power source. The 12-volt DC current is used to power the wide frequency range amplifier built into the antenna head. These antennas are directional and can be rotated by hand from inside the vehicle. Winegard also manufactures versions of this antenna for installing in homes or apartments. The company also makes indoor amplified antennas to improve OTA digital TV reception, as well as many other antenna styles.

Best OTA Antenna Types

For good OTA reception for far-away stations across a wide range of channels, try an LPDA (log periodic dipole array) antenna, which can be straight or V-shaped. Another choice is a Yagi, which has a lower threshold for tolerating or providing gain across a wide range of frequencies. Choose a Yagi antenna for the highest frequency of the band you intend to view.

A bolt-on accessory to improve steadiness and clarity of HD reception for the antenna shown in Figure 4-2 is the reflector assembly shown in Figure 4-3 (before installation). The installation is simple: Four plastic screw-hole covers are removed from the bottom of the antenna. The assembly is installed pointing away from the post by lining up the holes on the assembly with the four holes and inserting the

Figure 4-2 Amplified OTA antenna on an RV

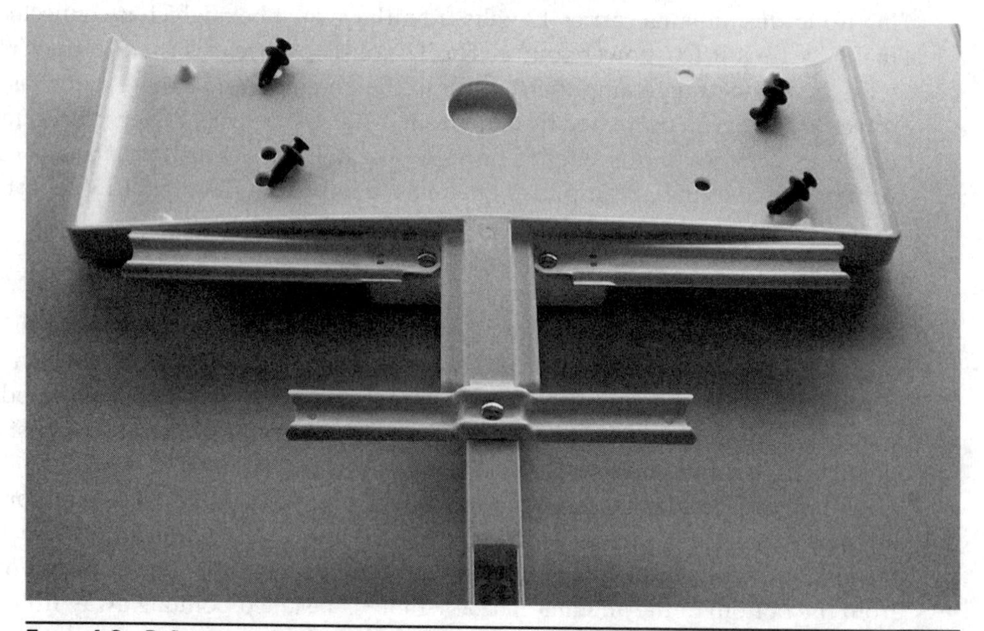

Figure 4-3 Reflector assembly improves digital/HD OTA reception.

Figure 4-4 Antenna with reflector assembly installed

twist-on plastic screws supplied in the package. The final assembly in use is shown in Figure 4-4.

How Satellite Dish Antennas Work

I could go through all the theory and mathematics to explain why a parabolic shape is necessary to get a sufficient signal to the FTA receiver, or I could demonstrate with something much more familiar to everyone. I'll choose the latter.

Light waves are simply the not-so-distant cousins of radio waves, and the two behave in similar ways. To help you understand the principle of the benefits of a parabolic dish antenna, you can perform the following experiment:

Shine a flashlight at a mirror. If you hold the lit flashlight perpendicular to the mirror's flat surface, a zero degree angle, the light will be reflected back exactly to its source. The physics theory rule regarding reflected light says that the angle of reflection is equal to the angle of incandescence. Test the rule by shining the flashlight at the mirror at a 45-degree angle (use your best judgment; no need to measure it). You will see that the light's reflection leaves the mirror at 45 degrees. Move the flashlight through an arc of 180 degrees and watch the angle of the reflected light change as you move the source. You can see the light reflected to opposite walls as you move through a full arc.

The radio (electromagnetic) waves from the satellite's transponder signal behave in a similar fashion when they strike the curved surface of the dish antenna. The radio waves approaching the dish are basically parallel to the dish and to each other, because of the huge distance they've traveled from the source. So the dish surface is hit by this energy field over its entire surface. Because of the curved shape, any energy hitting the surface of the dish is reflected to the same point above the dish's frontal surface. This point, when measured from the center of the dish, is called the *focal point*.

Regardless of where the electromagnetic field energy from a single wave originally strikes the dish surface, it is reflected and strikes the focal point at the same time. This happens because the total distance the reflected energy of that wave of energy traveled to get to the focal point is equal to the distance all of the other energy in that wave striking on the parabola traveled to get to the focal point. This is an important factor, because the wave's energy arrives at the collecting LNB in-phase, converging and creating a stronger signal. The phenomenon happens over and over again as each successive wave strikes the dish. This partially makes up for the energy lost over such great distances. The amount of gain for a given dish is a ratio of the energy received in relation to some point of reference and is expressed in decibels (dB).

Another use of a parabola shape is for audio listening devices, and it works in basically the same way. The sound strikes the surface, and the reflected sound waves are focused on the microphone located at the focal point. You might have seen parabolic listening devices advertised in comic books or in magazine classifieds. They collect so much sound energy traveling through the air because of their large size, and they focus that energy on a microphone mounted at the focal point of the parabola. Ever cup your hand behind your ear to hear better? You were using the same principle that makes a parabolic dish function, albeit in a more primitive way. So it is safe to say that the early application of the parabola shape is as old as mankind.

The larger the dish antenna (with a favorable geometry), the more wave energy will strike the dish surface and reflect back to the LNB mounted at the focal point, yielding a stronger signal to send down to the FTA receiver. Because of the dish's unique geometry, the collected electromagnetic wave energy reflected and received at the LNB from the dish's surface is in-phase regardless of where the energy in the wave strikes the dish's surface. This compounding of the energy in sync at the LNB creates enough of a voltage change to send energy down the cable to meet the receiver's need for processing the signal. There would simply not be enough energy to run the receiver from a 1-inch element, for example.

Types of Satellite Dish Antennas

Some variations occur in the design of satellite dishes and how they are implemented. In the next sections the text will discuss some of those variables relevant to FTA dishes.

Stationary Dish

A single stationary dish with one LNB can point at only one satellite to receive its signals.

Ganged Dishes

Two or more dishes can be separately aimed at separate satellites and the 4-to-1 switch can be used to select a particular dish as the channel is changed on the FTA receiver. These switches which allow the connection of one receiver to more than one LNB or dish are discussed in more detail in Chapter 8.

Motorized Dish

A single dish can be mounted on a pipe mast with a motor responding to a Universal Satellites Automatic Location System (USALS) command from the receiver, and the motor can rotate the dish to the correct satellite for that channel. The motor simply rotates the dish across an arc proportional to the Earth's horizon, allowing the dish to "see" all of the satellites on the horizon not blocked by a barrier or by the curvature of the Earth.

Portable Dish

Small, portable satellite dishes less than 30 inches (75 cm) in size are usually used with the pay-for-services satellite programming providers. A 30-inch dish mounted on a tripod is the smallest practical dish for an FTA portable.

Toroidal Satellite Dish

A toroidal dish is a unique design that uses two reflectors and multiple LNBs to acquire signals from multiple satellites across a 40-degree arc and a 60-degree azimuth change. Up to 16 LNBs can be installed on a 35-inch (90 cm) toroidal dish.

Automatic (Mobile) in Motion

In-motion antennas are motorized and will lock on to a satellite and maintain direction to that satellite as the vehicle moves down the road. These units work only with the pay-for-TV satellite providers.

Multiple LNB Dish

A dish with multiple LNBs mounted on the focal arm collects signals from satellites that are separated by only a few degrees. The parabola shape works with an offset angle to the original satellite and focuses the energy received from the second satellite to a new focal point, where a second or third LNB is mounted. Spacing of the target satellites and the extra LNBs cannot be much closer than 4 or 5 degrees.

You need to be aware of two tricks of the trade if you plan to experiment with multiple LNBs on the same dish:

- Set up the weakest satellite signal as the center one using the designed focal point of the dish. Place any secondary pickups to one or both sides of the center.

- Use two dishes with separations that skip what the first one is aimed at, maintaining a 4 degree or greater separation. The satellite signals that are closer than 4 or 5 degrees are set up to be received on the alternate dish.

It can be fun to experiment with how many LNBs can be received from only one or two dishes. Depending on your geographic location, the signal quality can be sufficient using this technique. At some point, though, the time and budget economics of a motor-driven dish prevail. Trying to receive from many satellites might be more fun than, say, bowling! For sure, it's more fun than shuffleboard.

Dish Assembly

The prototype selected for this section is a Pansat 90 cm (on the horizontal, as shown in Figure 4-5) dish. This dish will work well for receiving satellite signals during heavy cloud cover and can pick up some satellites that place a receiving location in the fringe areas of having only marginally sufficient EIRP (effective isotropic radiated power). Figure 4-6 shows the vertical axis measurement. For receiving FTA signals in the United States (lower 48), a 75 cm dish is considered sufficient by most vendors.

For FTA Ku band, you might choose a 75 cm, 90 cm, or 1.2 meter dish, depending on your available space, interests, or needs. In the Ku band, an efficient LNB on a 75 cm dish will receive signals that have an EIRP as low as 47 dBW. A 90 cm will push that number down to 44 dBW, and a 1.2 meter will receive signals as low as

FIGURE 4-5 Horizontal axis measurement on 90 cm dish in inches

FIGURE 4-6 Vertical axis measurement on 90 cm dish in inches

41 dBW. Looking at the satellite footprint maps for the signals you want to receive, your geographic location will yield the EIRP number. Use the EIRP to help you select a dish, or simply buy the biggest dish you can accommodate.

You don't need to know everything about a car to drive it, and you don't need to know everything there is to know about satellite dish antenna theory to assemble and use one to pick up FTA satellite programming. Between this chapter and the one-page assembly guide that might be included with your dish antenna, you should be able to assemble a dish in a few hours—or take a few days if you like or need to. The dish you acquire might be different from the one covered in this chapter, but very similar steps will be needed to assemble and mount it. Solicit some help if necessary to mount the dish and get ready to enjoy some free TV programming.

NOTE *One of the best pieces of advice I ever received was from an art instructor in a college art class in mixed flat media. I needed to cut along a line with scissors, and I found it difficult to cut exactly on the line. She told me to "look where you are going, not where you have been." This advice applied to art, and it has applied to just about every other undertaking in my life. Cutting along a line or reaching a major goal became a lot easier from then on. So now it's my turn to advise the reader. When you are following step-by-step instruc-*

tions, after the first step, always read them in groups of three. Read the step you just performed, the step you are going to do now, and the one you will do next. As you work through this dish assembly project, keep the mental spotlight on the moving three instructions sets and the assembly will go much better. It takes more time to check the previous steps as you go, but so does going many steps back to correct mistakes.

Tools

You'll need the following tools to assemble this dish:

- Flat-blade long shank screwdriver
- 10 mm open end/box-end wrench
- 7/16-inch open end/box-end wrench
- Phillips #2 screwdriver
- Electric drill motor
- 13/64-inch drill bit
- Pencil
- Ruler
- Level

The dish arrives in a large cardboard carton, and the adventure begins.

Assembly of Dish, Feedhorn (LNB), and Adjustment Flange

1. Remove the outer carton by cutting the flaps on one end and lifting the opposite end to reveal the components. This side is up:

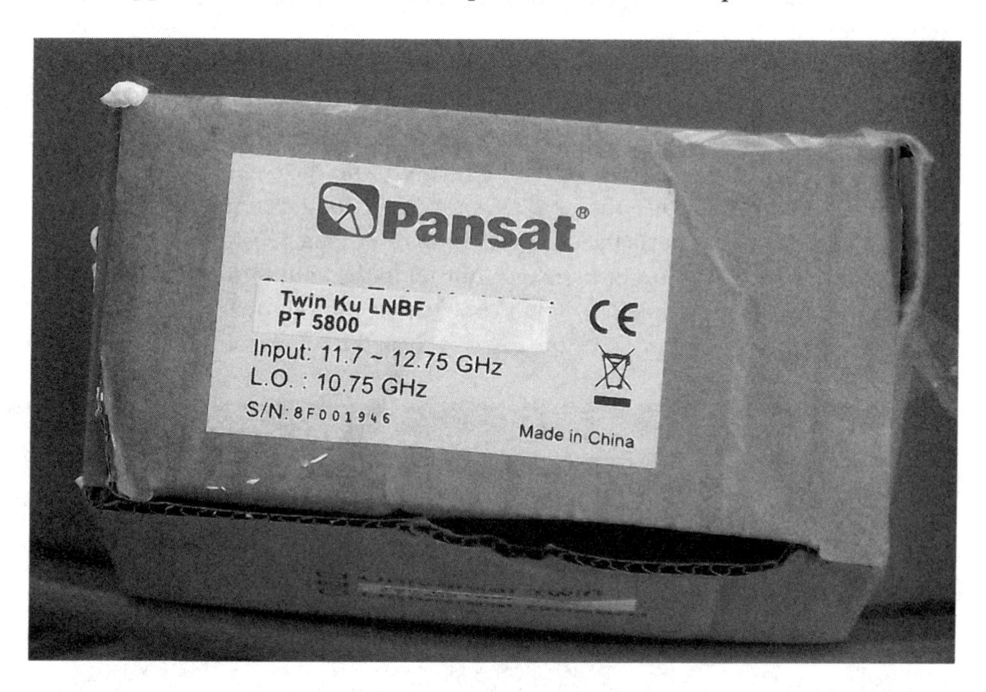

2. Retrieve the LNB feedhorn box and set it aside in a safe place. The LNB is the most fragile of the parts needed for the dish antenna. It will be installed toward the end of the assembly.

3. Break out all the pieces and parts, lay them out, and match them to the bill of materials included in the box. It would be unusual to be missing a part, but it can happen. At this point, leave the hardware, nuts, machine screws, and washers in their respective bags. Do not open the bags until it's time to use the hardware.

   The large curved pipe is the mounting mast. The larger diameter braces will help hold the mounting mast in a vertical orientation. The braces connect to the toothed circular clamp and to the mounting surface for the dish. The L-shaped brace will hold the LNB, and the two narrow diameter braces will help steady the LNB's L-brace. The longer of the two 90-degree offset flanges connects the dish to the mounting mast, commonly called the "butterfly." This larger flange will provide the ability to adjust the dish for elevation. The smaller flange connects to the bottom of the mast and to the mounting surface for the dish, such as the porch roof or pole.

4. The actual assembly could begin at any number of points. For the first-time assembler, it is easier to start by attaching the butterfly to the back of the dish. At this point, open the bag and look for two bags of hardware and the mast pipe clamp. One bag contains mostly machine screws, hex nuts, and washers for dish assembly, and the other includes six large lag screws for attaching the smaller flange to the house or pole.

5. Empty the dish hardware bag to separate and lay out the hardware pieces, matching like items to like items. Notice the machine screws are of a par-

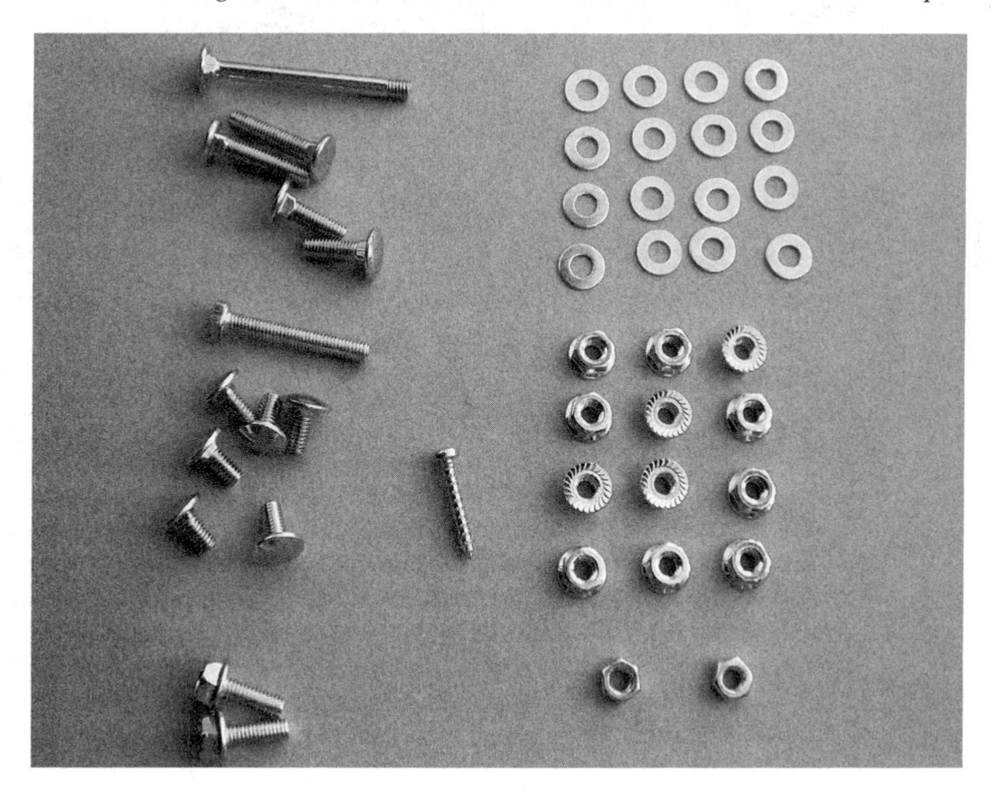

ticular type, with a square neck above the thread and in front of the rounded head. These are called *carriage bolts*. If you examine the clamp, the dish itself, and the flanges, you will notice many square holes punched into the metal. The square neck on the carriage bolts fits in the square holes and allows tightening of the carriage bolt with a wrench at the nut end only. The rounded heads of the carriage bolts are smooth and present a low profile where clearances are an issue. Two of the bolts are carriage bolts with completely flat heads and are used under the butterfly to attach the L-brace.

Notice the four square holes near the center of the dish. These holes are for mounting the butterfly flange on the back of the dish, using four each of the shorter carriage bolts, serrated flange nuts, and flat washers. For reference, the top of the dish has the manufacturer's name on it.

6. Take a close look at the butterfly flange and the pipe clamp. The clamp is reversible so that the orientation of the face of the dish can be adjusted for vertical angle. Sides A (square hole A to round hole A) together will allow dish angles from the teens and up to and into the 60-degree range.

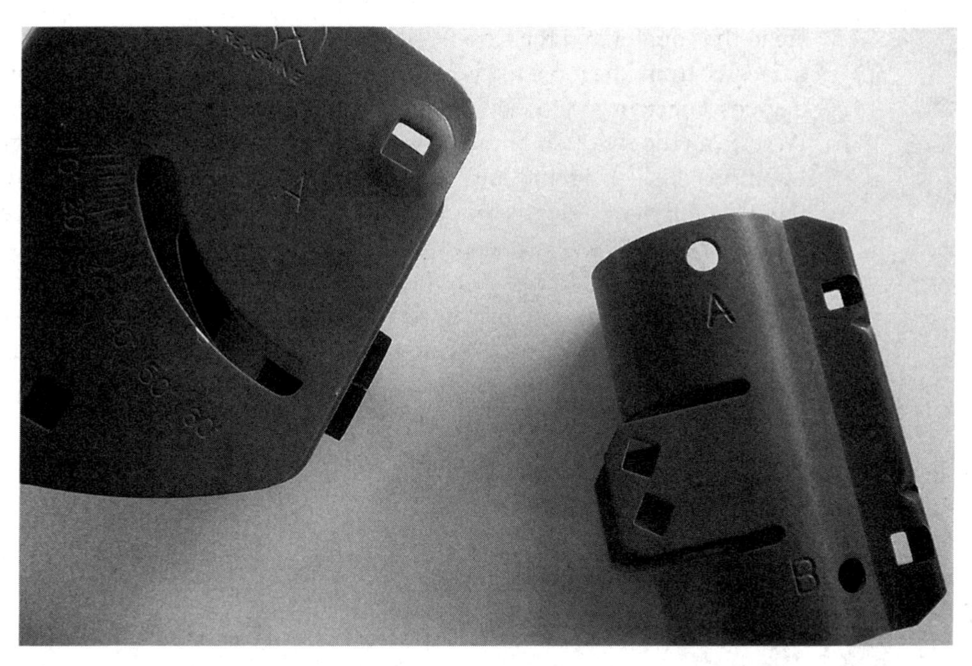

Flipping the orientation of the butterfly flange and the pipe clamp to sides B permits dish orientations to the vertical from 50 to 90 degrees. At this point, you need to know what orientations for vertical alignment will be necessary from your location on the planet. As a general rule, the closer you

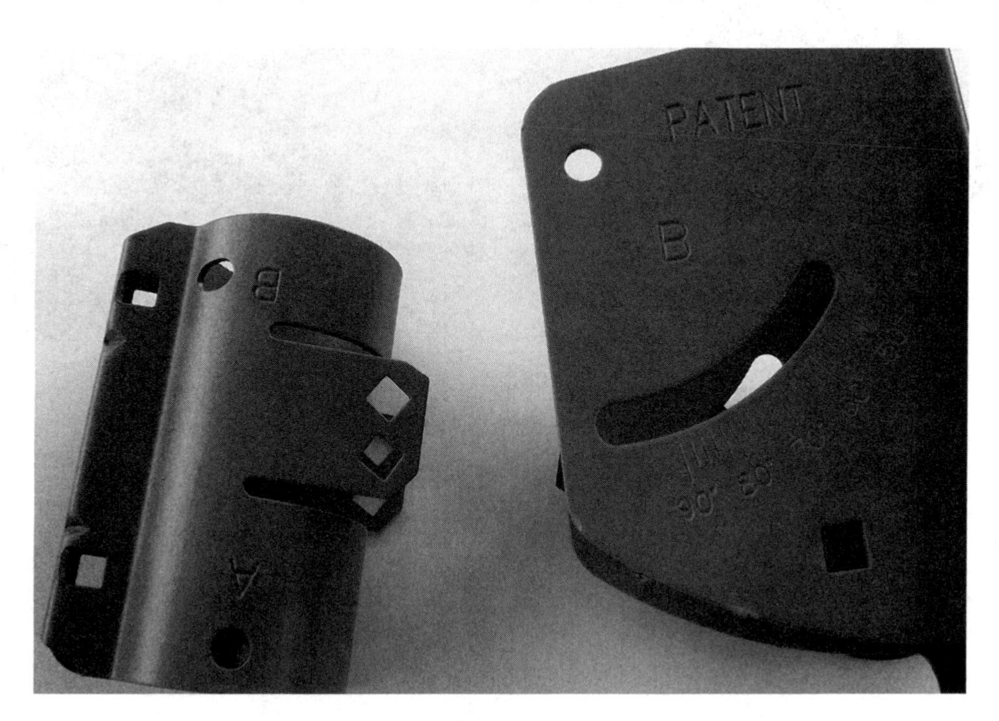

are to the equator and the closer you are in latitude to the satellite you want to receive from, the more likely it will be necessary for orientations above 60 degrees. For example, to view programming from Horizons 2 at 74.0 degrees West from Bogota, Columbia, the dish elevation will be 84.6 degrees. Locations such as Miami and Key West, Florida, and Brownsville, Texas, will reach above 60-degree orientations for some satellites. Most places in the lower 48 will have dish elevation orientations below 60 degrees.

7. Slip the pipe clamp in between the flanges of the butterfly flange, making sure that the correct holes are lined up for the range of elevations needed. Because this assembly is for locations north of the equator, check to make sure that the long machine screw will be placed in square hole A on the butterfly flange and then through round hole A on the pipe clamp. This way, the finished assembly will work for adjusting the dish's elevations from in the teens of degrees through and into the 60s of degrees.

8. Insert the long carriage bolt through the square hole labeled A on the butterfly flange and through the round hole A on the pipe clamp. When you are finished, it will look like this:

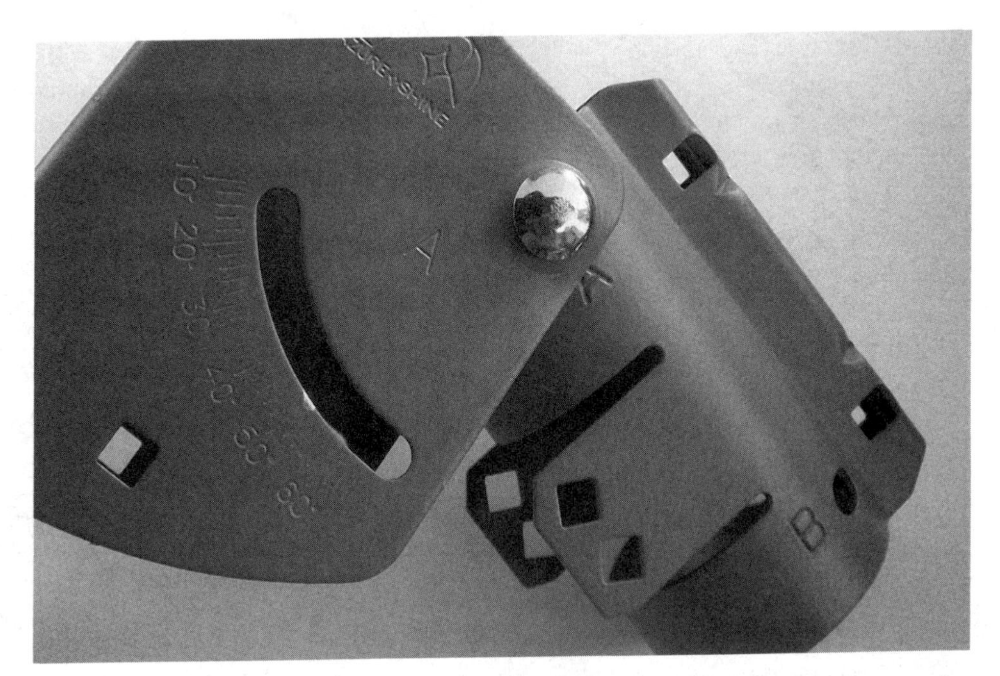

On the other side is a flanged nut and washer. Do not put washers on the bolt head side when using carriage bolts. When you are assembling metal parts, do not tighten the nuts and bolts beyond "finger loose" until all the pieces are together.

NOTE *By "finger loose," I mean turn the nuts down just until you feel the resistance of the metal, and then stop turning. "Finger tight" is a finished condition, where you torque a nut, such as a wing nut, as tight as you can by using only fingers/hands. Rotate nuts right for tight and left for loose.*

TIP *You can upgrade the elevation adjusting part of the butterfly flange using wing nuts with the three carriage bolts used to clamp down the elevation setting. If you decide to do that on this unit, remember the threads are metric sizes. Using wing nuts makes frequent elevation adjustments easier.*

9. The pipe clamp at this point in the assembly should swing freely inside the flange. Three carriage bolts will secure the elevation setting on the finished dish assembly. With the long bolt in place, swing the pipe clamp into the flanges until the arced slot of the flange aligns with one set of the square holes in the pipe clamp, one on each side.
10. From the inside of the clamp, insert one of the small carriage bolts through the clamp's square hole into the arced flange slot, placing a flat washer and serrated flange nut on the bolt and tightening slightly. The illustration shows the second carriage bolt ready to be inserted on the opposite side.

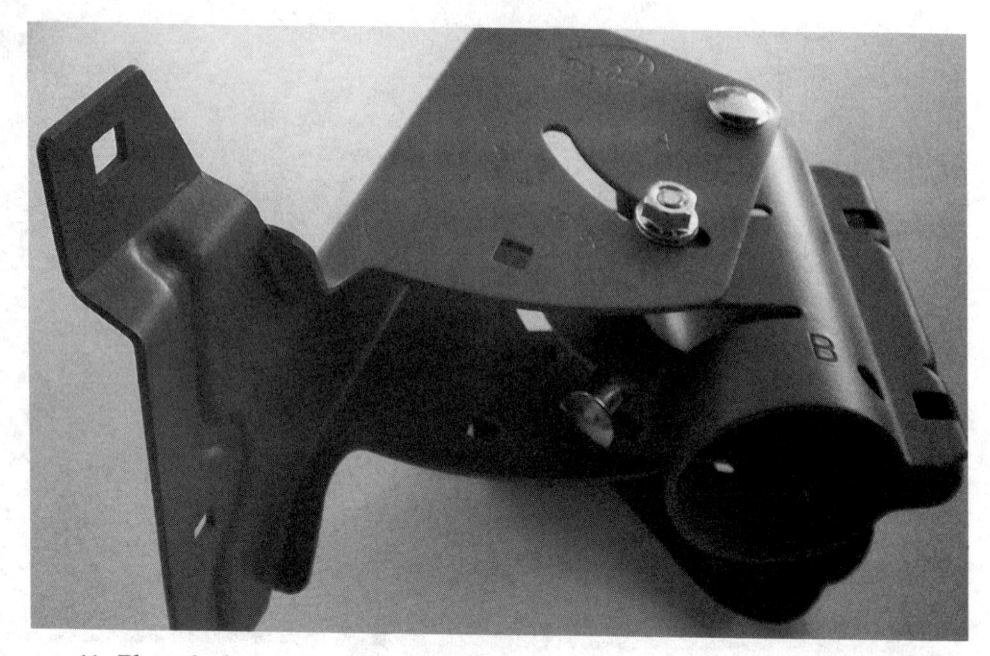

11. Place the butterfly flange with the pipe clamp attached to it on the back face (convex surface) of the dish near the four square holes. Notice that an arrow is cut into one wing of the butterfly flange. This arrow must point to the top of the dish. Because the holes are offset, to install the dish correctly, the arrow must be pointing toward the top. Look at the other side of the dish

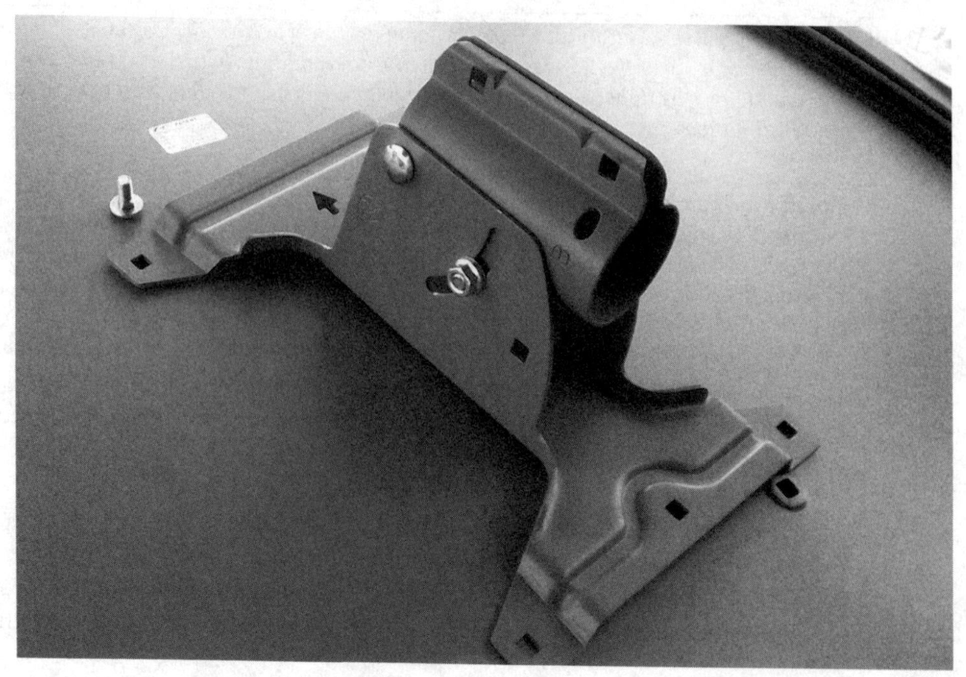

and read the manufacturer's label information, if necessary, to determine where the top of this dish is. At this point, it's nice to have a workbench or table, long arms, or a helper to put the first short carriage bolt into the upper-right square hole from the concave side of the dish. After that is done, place a flat washer and flange nut very loosely on the bolt, with perhaps one revolution of the nut—just enough so the butterfly flange will stay connected to the back of the dish. Final tightening will be done after all assembly steps are completed.

Viewing from the right side, notice the one nut in the upper-right holes of the dish and butterfly flange. Notice that the arrow is pointing to the top of the dish. Many parts of this assembly can be done in a sequence different than the instructions here. This next step, however, must occur before any more of the butterfly connector bolts are installed.

12. Find the two carriage bolts whose heads are different from all of the others. They have flat heads rather than rounded, because of the low clearance between the butterfly clamp and the back of the dish. Using the wrong bolt could damage the surface of the dish and change its electrical reflective properties. Lift the butterfly flange away from the surface of the dish and install one flat-head bolt up through the square hole in the bottom of the flange.

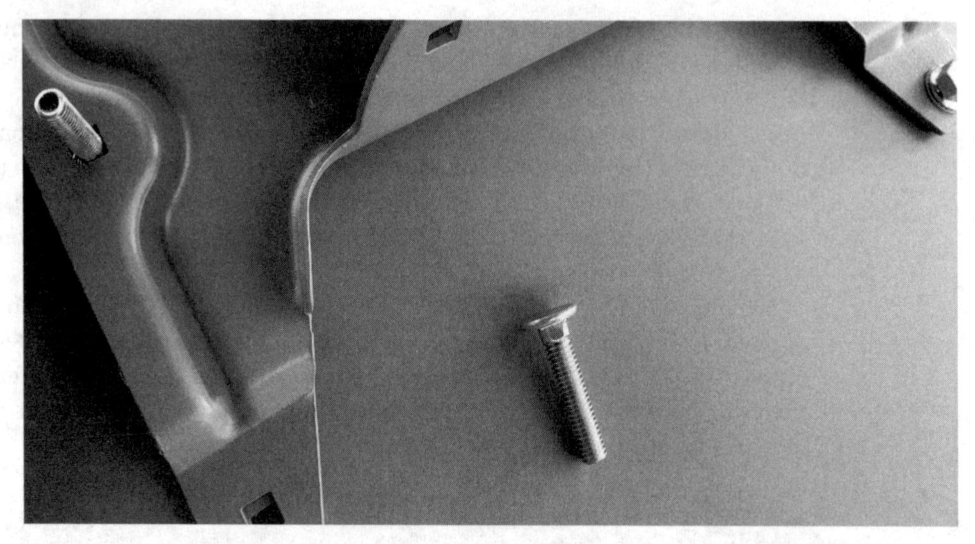

13. While holding the flat-head carriage bolt in the butterfly flange, lift the flange away from the dish surface again and insert the second flat-head bolt in the center (the next available) square hole, leaving the top one empty. Then return the flange to the surface of the dish, lining up the remaining three mounting holes so the dish can be securely mounted to the butterfly flange with short carriage bolts. Check to make sure everything is lined up, as shown in the next illustration.

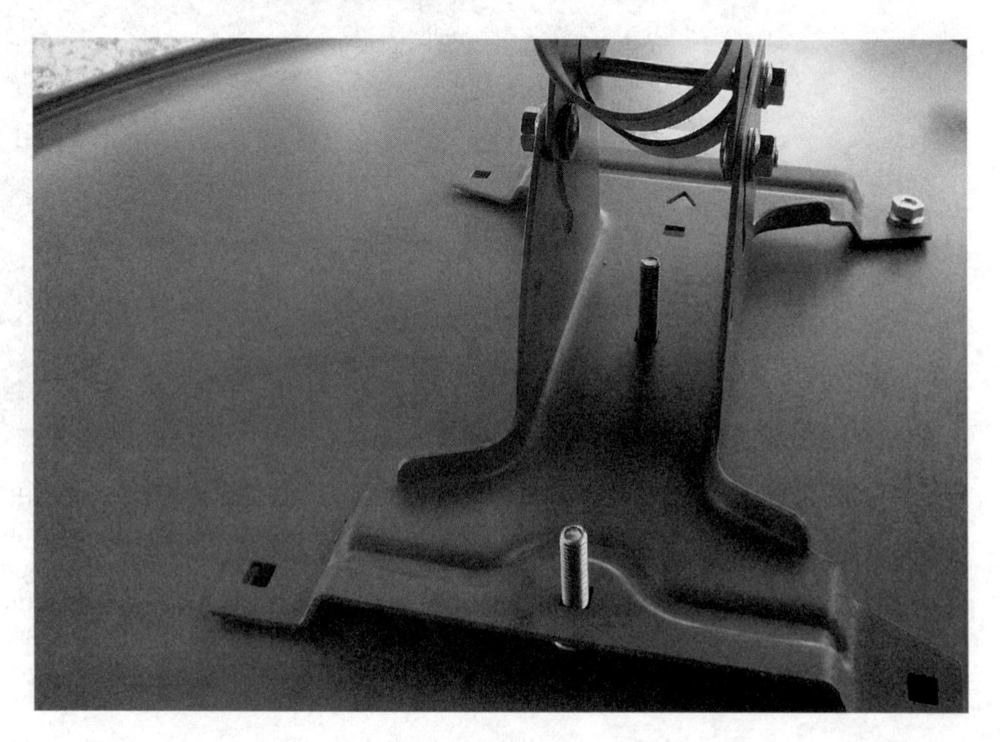

14. Insert the remaining three short carriage bolts, one at a time, through from the front of the dish into the butterfly flange, and place the flat washers and flange nuts on each one as it is inserted. Loosely tighten the nuts on all four bolts. Check the front to be sure the square shoulders of the bolts are through the square holes in the front of the dish and that the bolt heads are flush with the inner surface of the dish.
15. The L-shaped brace for the LNB is the next piece to install. Place the bottom of the dish over the edge of a table or workbench to make this step easier if it is not there already. This step will require use of a long-shanked flat blade screwdriver. Insert the screwdriver under the flat-head carriage bolt to push it up, and then place the L-shaped brace over the bolt to line up with the third hole from the top of the L-brace. Place a flat washer and flange nut on the bolt and tighten loosely. At this point, the foot of the L extends out and under to the front side of the dish to suspend the LNB at the focal point of the dish.

16. The second flat-head bolt should already be lined up into the top hole of the L-brace. Place the screwdriver under the second flat-head bolt and attach the washer and loosely tighten the flange nut.

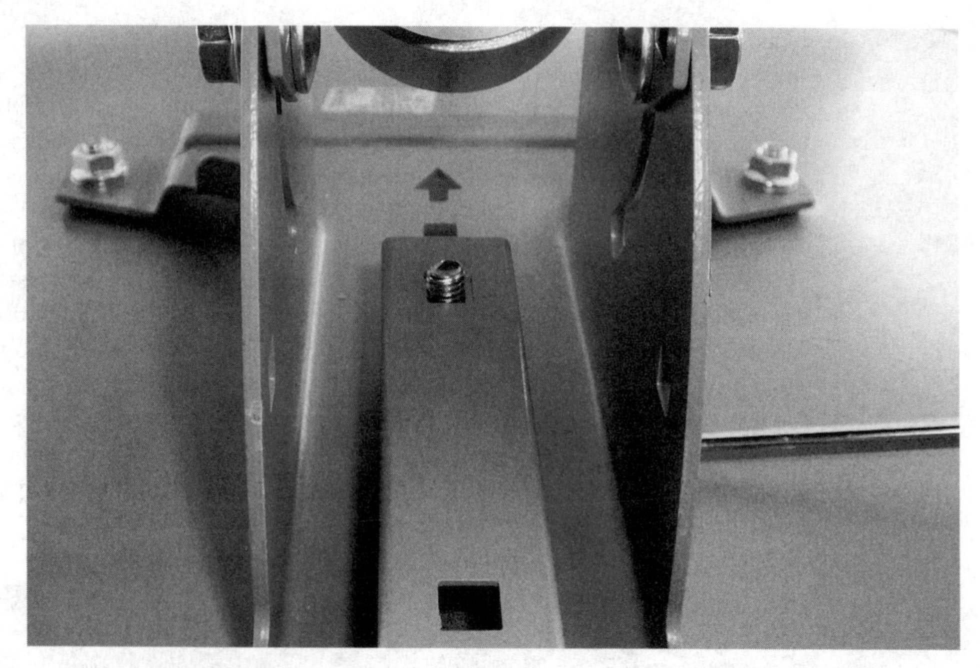

17. Install two more each—carriage bolts, washers, and flange nuts—on the pipe clamp for the eventual insertion of the mounting mast pipe.

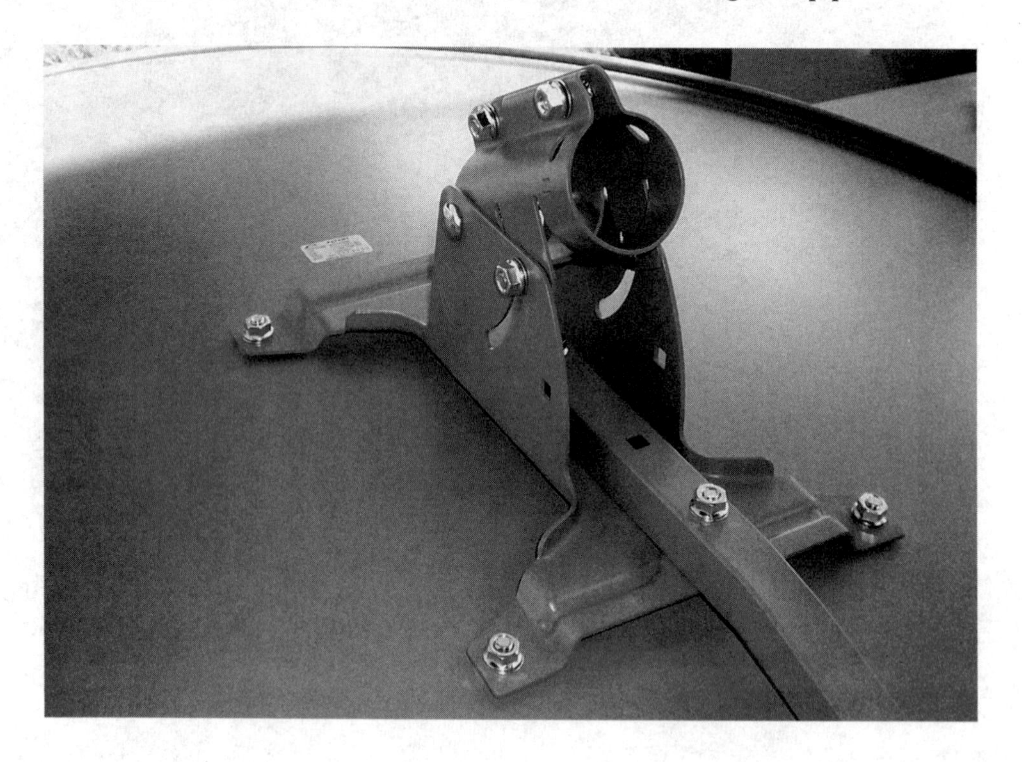

18. Carefully position the mounting mast to be inserted into the pipe clamp. The mast mounting pipe has a bend at the bottom for the style of installation in these instructions. Insert the long end after the bend into the pipe clamp on the butterfly assembly. The top of the mast pipe should be just under the long bolt passing through the clamp when the pipe is inserted, as shown in the next illustration.

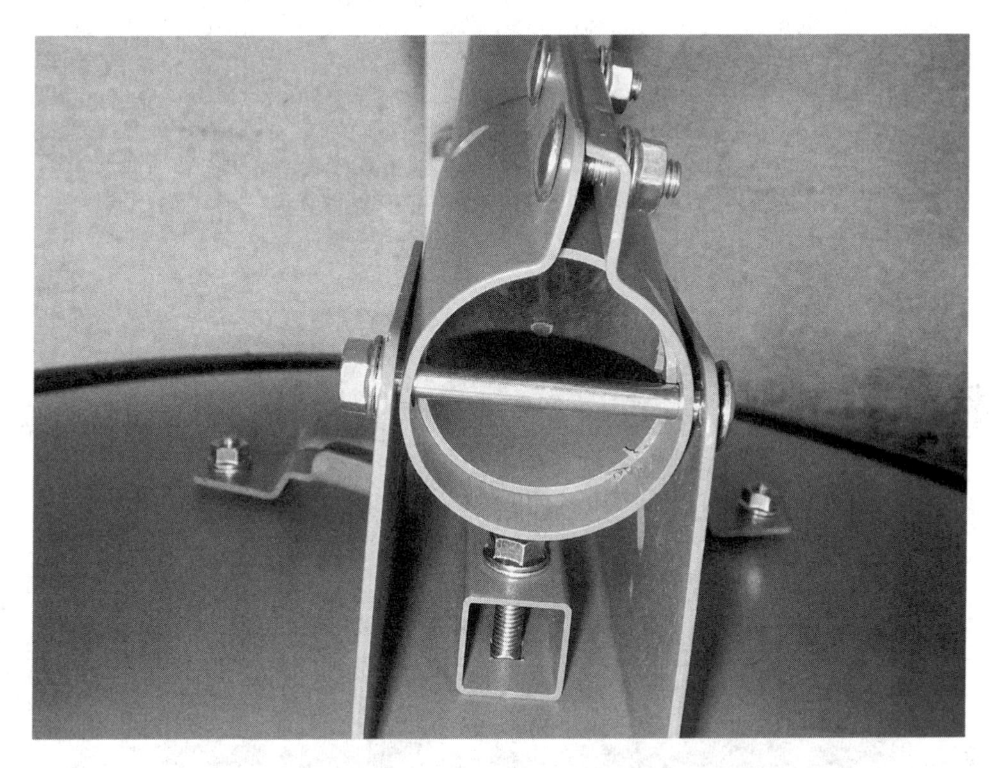

19. Go back to the breakout illustration for step three and notice the mounting mast pipe has the same holes drilled and punched in each end. The reason for this is that on some final installations, the pipe will be reversed end-for-end using the longer throw of the pipe to reach under a roof eave, for example. Assembly instructions here are based on ground mounting on a flat surface. Having the through-bolt (fulcrum bolt) ride on the top of the pipe is necessary for adjusting the direction later as far as the satellite look azimuth is concerned. This allows the dish to be rotated manually through azimuth adjustments of some number less than 360 degrees on the mounting post, depending on the full geometry of the installation. The two pipe clamp bolts are more than capable of holding the dish to a correct azimuth when fully tightened after final azimuth adjustment. The down-lead cable and the LNB arm clearance and dish clearance are the delimiters for full rotations. This would not matter so much for mounting in portable mode where the dish

(tripod or other mounting hardware) can be rotated for dish azimuth adjustments. In that case, it would be OK, but not necessary, to remove the bolt and insert it into the top hole in the pipe for added stability.

The second alternative at this point is to install an automated positioning motor. Install the positioning motor according to the manufacturer's instructions between the mounting pipe mast and the dish. Many variations are possible, so follow the manufacturer's instructions. The motor must mount to a correct mast pipe size.

20. From the bottom of the mast pipe, slip the toothed clamp over the pipe and slide it up to the bottom of the pipe clamp. It will need to be in place for connection later. Consider putting a piece of masking tape or transparent tape temporarily on the clamp so it won't slip off.

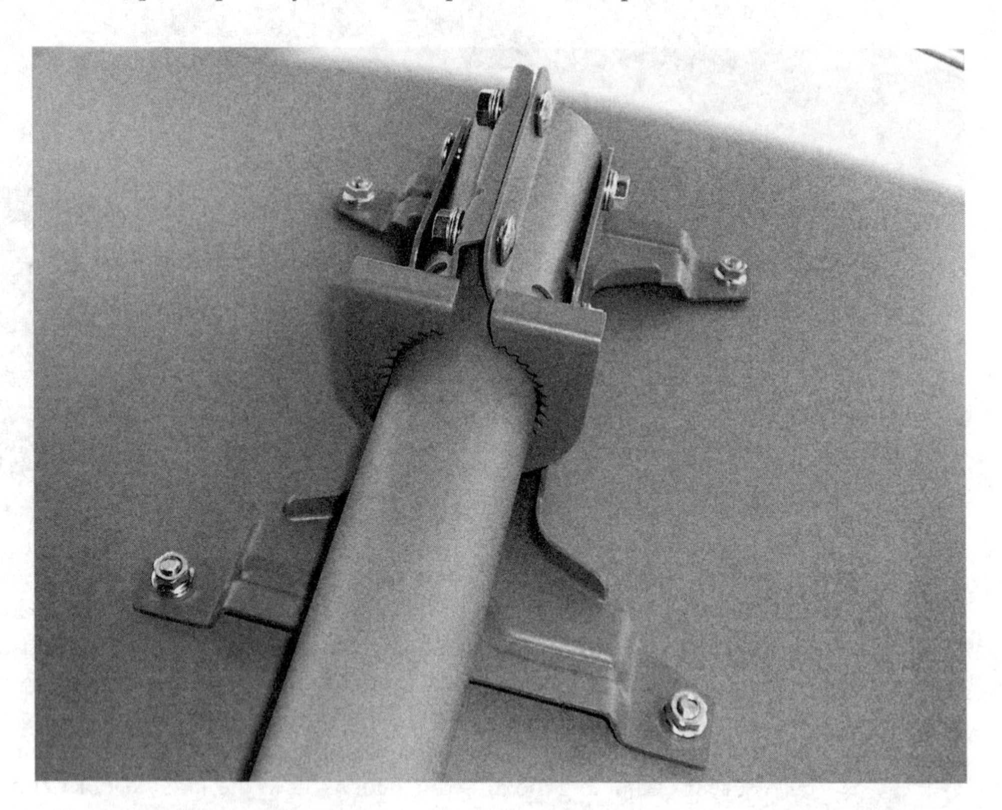

21. At this point, the focus of the assembly will move to setting the braces in place to mount the LNB L-brace and holder securely. Carefully flip over the disk assembly so it is resting on the back of the dish with the mast pipe on the work surface or bench. Find the LNB brace with the letter *R* stamped near one end. Use the standard machine bolt, ordinary hex nut, and flat washer to connect the brace to the raised rim of the dish.

As for the dish being right or left, think like port or starboard on a ship, which never changes. With the dish pointing at the sky, and you behind the dish, the right side of the dish is on your right. If you move to the front of the dish, facing the dish the right side of the dish is still the same. Bends in the ends of the brackets will work only if the R-stamped end is installed on the right side of the dish and the L-stamped brace is installed on the raised rim on the left side of the dish.

22. After the round hole near the R is lined up with the hole on the raised rim of the dish, insert the bolt with the flat washer from the front going through the brace and then into the round hole in the dish. Place the nut on from under the dish. Do not use a washer under the nut. The raised rim of the dish will mesh against two flats of the hex nut and keep it firmly in place. The flanged bolt will be tightened later with a Phillips screwdriver. So for now, tighten just enough to get the nut to slip into the underside of the raised rim and for the brace to stand up at its designed angle to mate up with the LNB plastic bracket.

The attachment of the first LNB brace is a good point to go and find the LNB and the LNB mounting bracket. With the first brace attached, the assembly should look like the illustration on the next page.

23. Pass the medium-length hex head bolt with flat washer attached through the bracket, the hole in the plastic LNB bracket, and into and through the L-brace. Notice a hook and eye at the top of the LNB bracket and an alignment pin just below the semicircle.

24. Install the second brace that will help support the LNB. It will be marked with an *L* for the left side of the dish. Insert the bolt with the flat washer from the front going through the brace and then into the round hole in the dish. Place the nut on from under-side of the dish again without a washer.

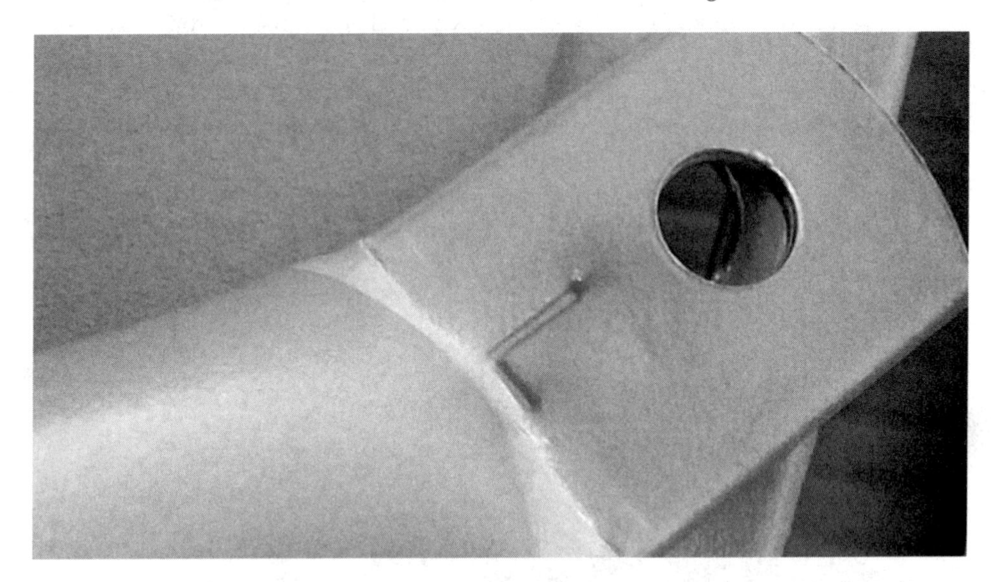

25. Now install the LNB into the two halves of the plastic bracket and secure the brace.

26. Place the opposite half of the plastic LNB bracket around the LNB, and then clip the hook and eye into each other. Swing the right half into place to line up the alignment pin in the hole for it on the opposite bracket. The LNB is installed with the round microphone-like end pointing at the dish.
27. Place the square notch in the LNB bracket against the top of the L-brace so that the bolt can continue through from the left side bracket half, through the L-brace, and through the right side bracket half. The LNB should be relatively loose inside of the bracket and should rotate freely.

NOTE *This assembly step is a bit tricky and requires that you use both hands, as six parts are involved in a process that brings them all together. It's a bit tricky at times to get all these pieces to come together. Be assured they will come together and without forcing anything. Be particularly careful not to damage the plastic bracket or the LNB. It's OK to remove the left bracket momentarily and practice this assembly away from the L-brace before trying to pull all the pieces together.*

With the LNB swung to one side, look between the halves of the plastic bracket to make sure that the alignment pin is lined up and in the hole in the opposite bracket.

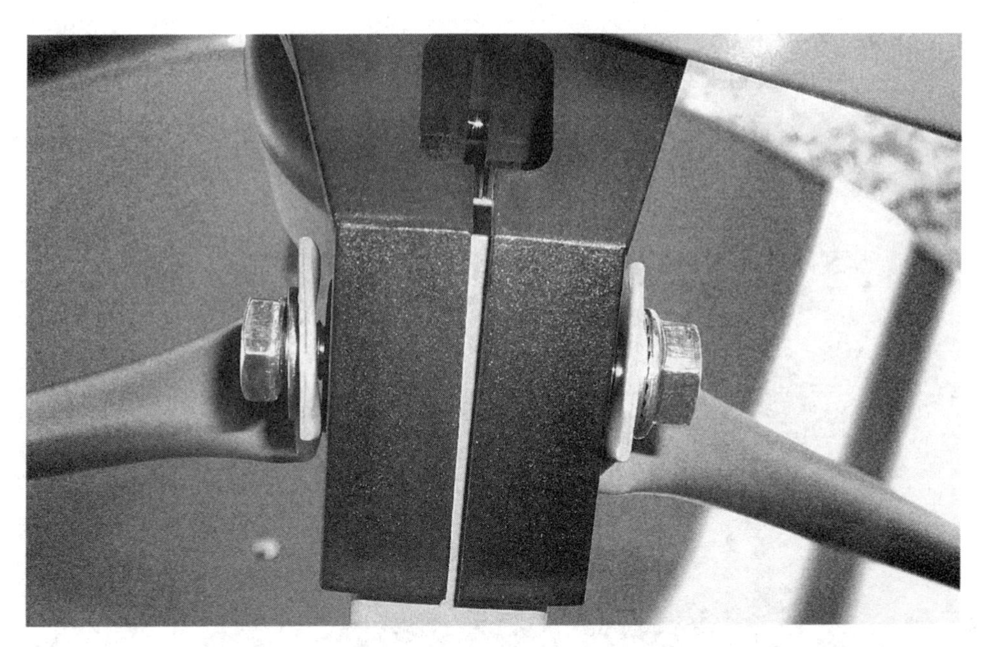

28. The final step for LNB mounting is to install the Phillips-head screw that will be used to lock the LNB firmly in place. Notice that the screw hole is just below the circular halves of the bracket.

NOTE *After lining up with a satellite later, some adjustment of the LNB might be necessary. Chapter 7 discusses dish alignments and fine tuning.*

Notice the label on the bottom of the LNB. It shows the technical data about the LNB frequency domains. The label says "twin" LNB, which means it has outputs for connecting cables to two receivers.

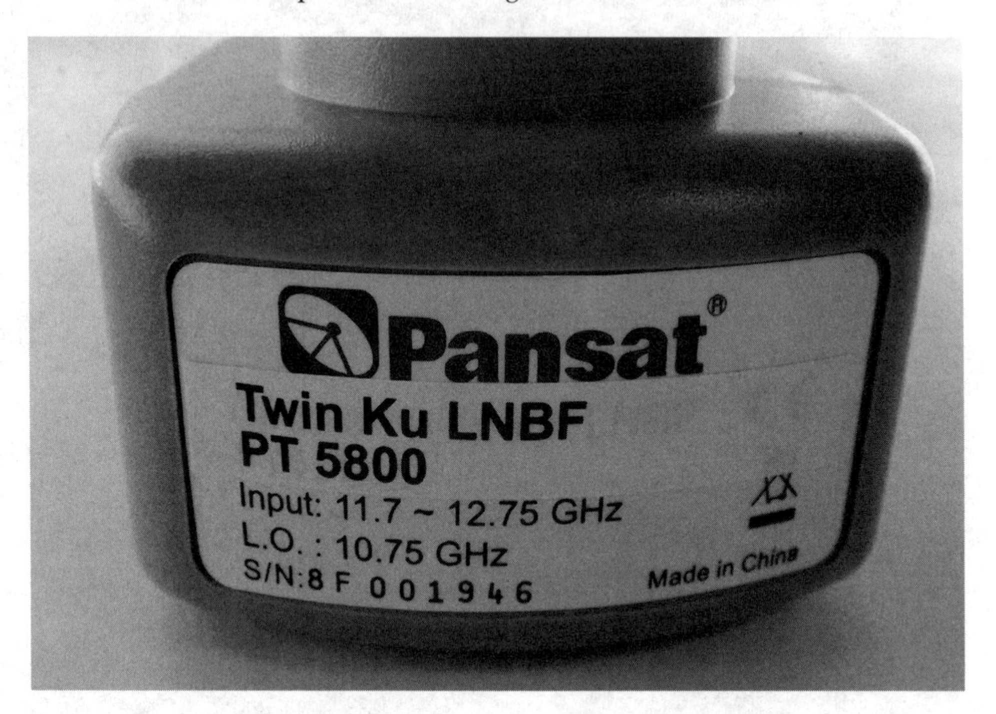

Mounting the Flange, Toothed Clamp, and Braces

Now we'll assemble the mounting flange, the toothed clamp, and the mounting braces. This universal mounting bracket and hardware provides a plethora of options for securing the dish to a permanent or semi-permanent mounting place. Mounting to concrete, the solid wood of a structure, a concrete block foundation of a building, on a pole or post, or on a sturdy fence are all possibilities. Select a sturdy option for mounting, because the dish acts as a sail in the wind and might have to withstand wind velocities in excess of 100 mph (160 kph).

1. Open the second hardware bag and lay out the remaining parts, as shown in the next illustration.

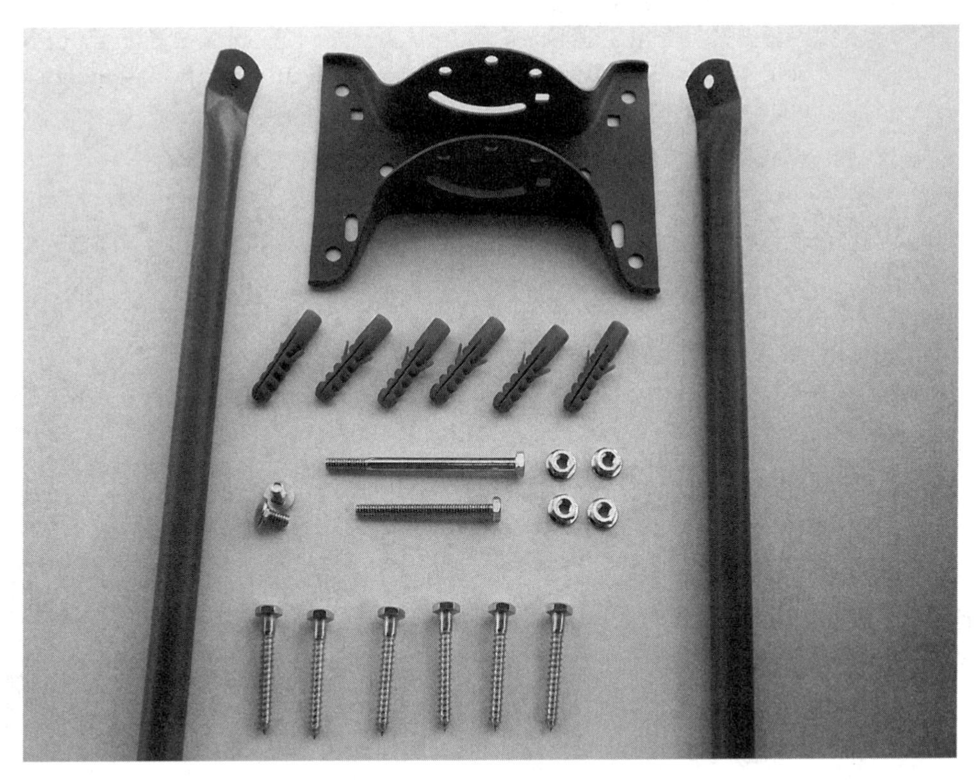

2. Attach the left side brace to the toothed clamp by inserting the long-shanked machine screw through the left brace into the toothed clamp.

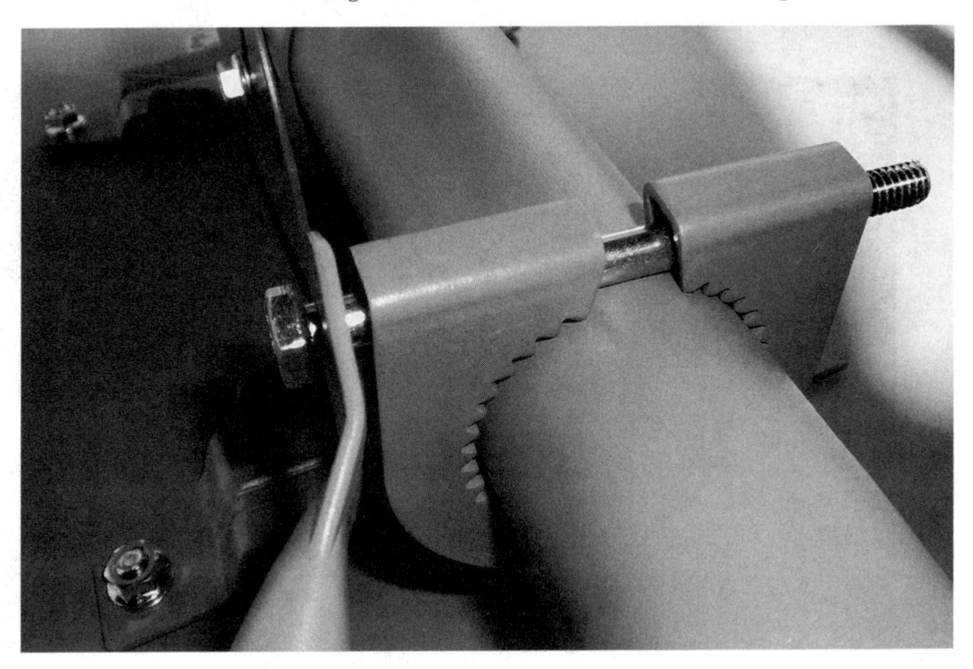

3. Attach the right side brace and place a flat washer and flange nut on the bolt. Tighten loosely. Again it is best to wait until all the assembly is complete before totally tightening all the pieces together.

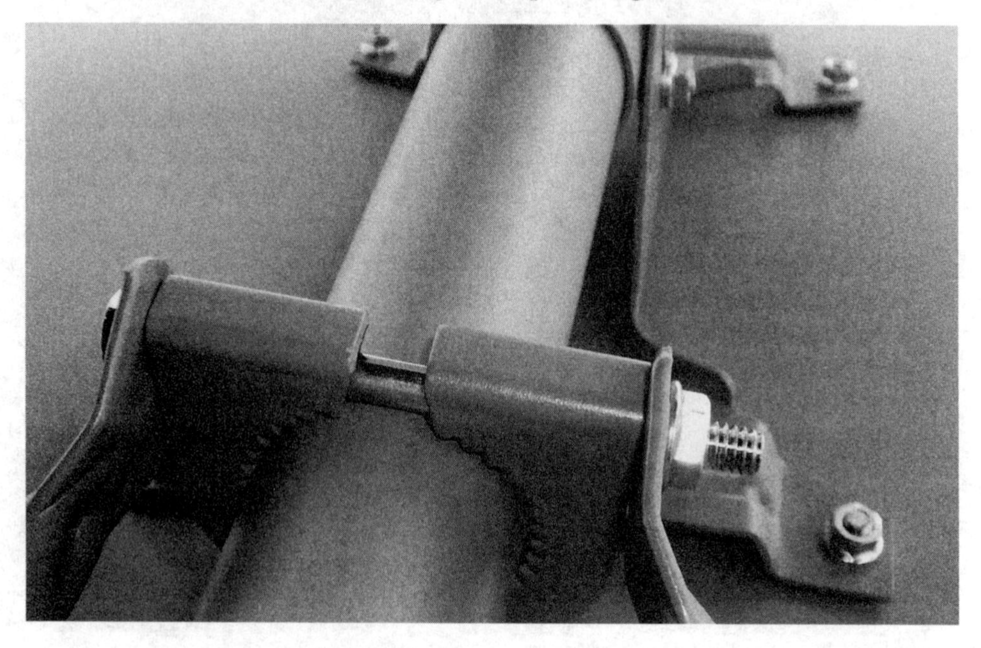

4. Now you'll hang the mounting flange on the bottom of the pipe. Place one of the small carriage bolts through the square holes in the bottom of the pipe from the inside of the pipe.

The carriage bolt will slip through the arc slot in the mounting flange, as shown in the next illustration.

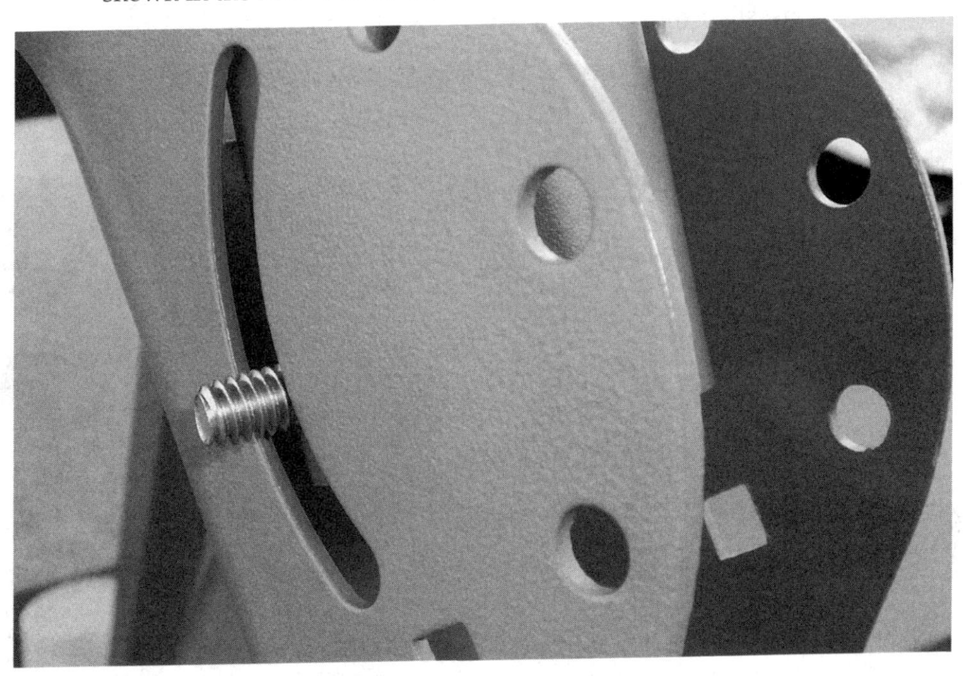

5. Place the second carriage bolt from the inside of the pipe through the arc slot on the other side of the mounting clamp.

6. At this point, the mounting flange will just hang on the pipe so that the flanged nuts can be installed. No washers are necessary, and again tighten just slightly.

7. The bracket is a universal bracket with plenty of angle swing and extra holes, allowing for mounting in any number of orientations from its mooring place. For now, assume this will be mounted in a vertical orientation of the mounting mast, so we will use the center hole as the fulcrum point.

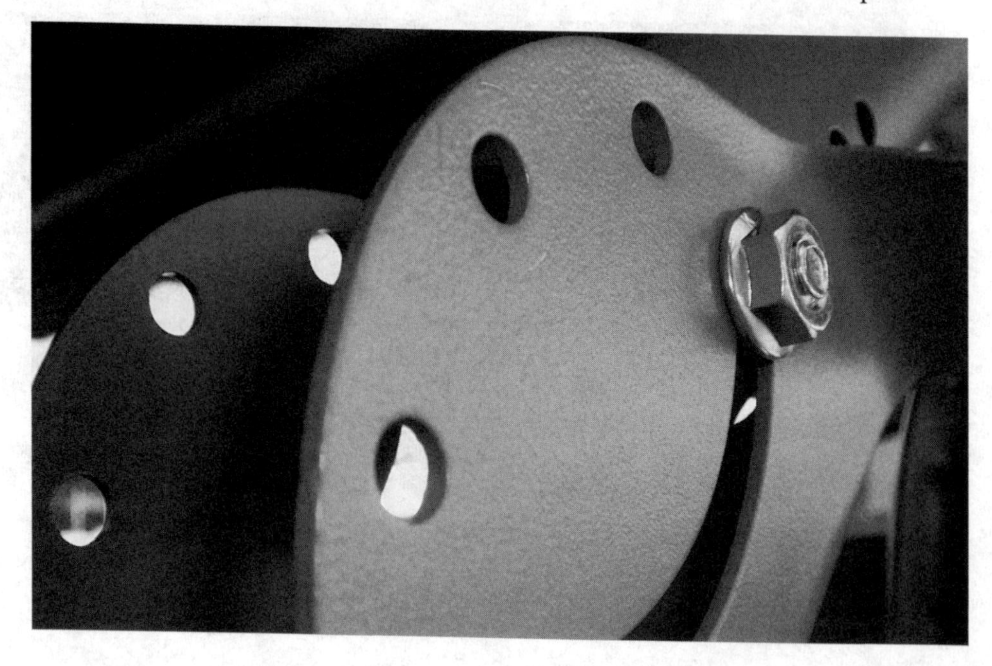

Swing the mounting flange up so that the center hole in the flange lines up with a hole in the mounting pipe when the flange's foot is perpendicular to the pipe. Use the hex-head fully threaded bolt, the smaller of the two from the package, with a flanged nut. No washer is needed on either side. When the dish and flange are permanently mounted for use, this bolt will be torqued very tight and the serrations on the flanged nut should maintain the friction connection. The braces will add extra stability.

Notice twelve holes on the bottom (foot) of the mounting flange that are used for permanent mounting. At a minimum, use the four corner holes for securing to the mounting surface. Two holes are square and provide an opportunity to place a flat-head carriage bolt pointing up for securing something to the flange without your having to fit a wrench underneath the flange. For example, this could be used to secure a guy wire for an overhead cable run or be used during the final mounting steps to secure the down-lead cable with a clamp. Square holes are repeated on the side of the flange for the same reasons.

The next illustration shows the mounting flange installed perpendicular to the mounting mast pipe. The two other holes are for mounting the pipe at offset angles to accommodate a variety of angled mounting surfaces.

Of the supplied hardware, all that should be remaining are six lag screws and corresponding concrete anchors.

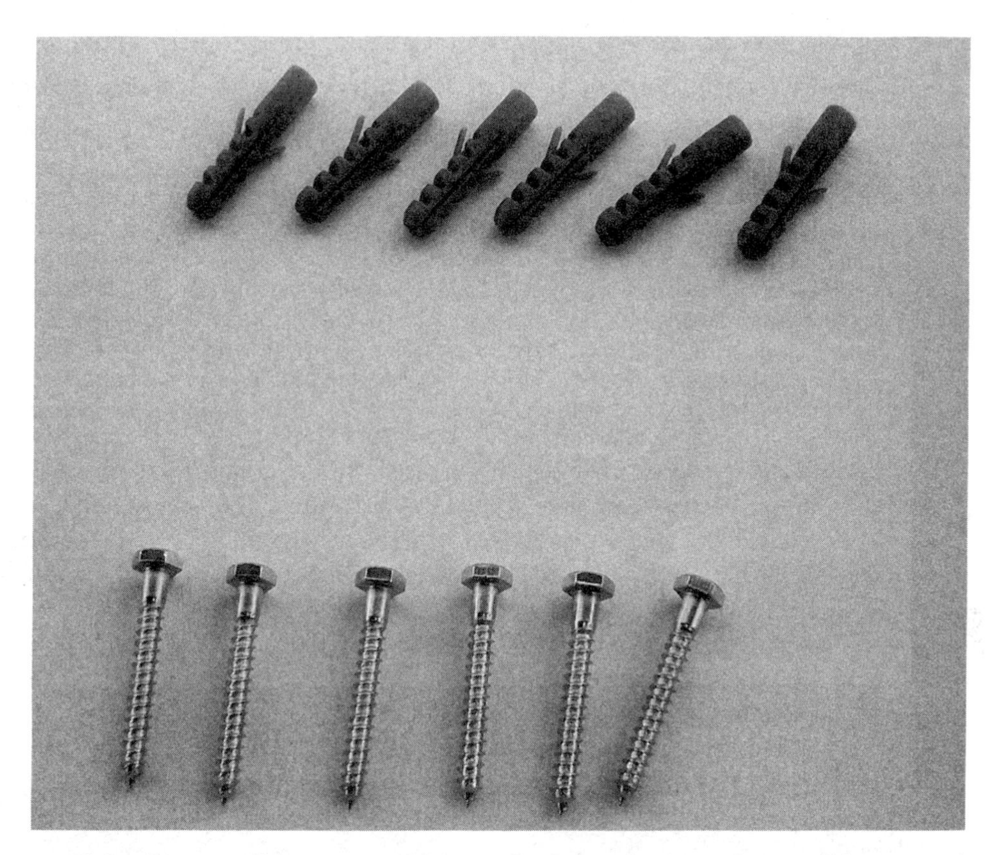

If the flange will be mounted in concrete, the plastic anchors will be pressed into drilled holes to receive the lag screws. The plastic anchors expand when the lag screws are tightened, causing the plastic anchors to grip the walls of the predrilled holes. Drill anchor holes in concrete the same size as the anchor's diameter, or follow the recommendations of the anchor manufacturer when buying additional mounting hardware.

If you are mounting in solid wood, the plastic anchors are not used and the lag screws are set in directly and tightened in smaller pilot holes. The rule of thumb I follow for wood pilot holes is to drill a pilot hole no bigger than three-fourths the size of the diameter of the screw in hardwoods (prone to cracking) and no smaller than five-eighths the diameter in soft woods. As for drill depth, I drill the pilot hole to three-fourths of the length of the bolt and at least ½ inch less than the length of the screw. Look at the recommendations of the screw manufacturer for specifics; they are often included on the packaging from the hardware store.

NOTE *The worst case scenario is that the lag screw snaps in two during final torqueing and you have to start all over. Also, if the pilot holes are drilled too big, the dish can blow away in the wind. Take care to provide a sound and secure mounting for the dish.*

In the next sections, the mounting mast pipe and braces will be placed on a temporary stand to illustrate steps needed for mounting the dish on a horizontal surface.

NOTE *I am a big advocate of using ground-mounting opportunities whenever possible. I use an existing in-ground concrete surface, a basement wall, or a sturdy fence, or I sink a short pole in the ground or pour a 3-by-3 foot or 4-by-4 foot pad—anything that keeps me off of a ladder. The added advantage is that a little readjustment later is easier, because you can walk to the dish, and you don't need a ladder. Ground mounting, or near ground mounting is not possible without a clear unobstructed view of the sky and not advised in areas prone to vandalism. Survey your site to determine what view of the southern sky (or northern sky if in the Southern Hemisphere) will be best in your unique circumstances.*

1. Trace the flange with a pencil to identify the four corners for mounting holes. Alternatively, you can make a cardboard cutout, marking the holes if the installation is above ground level. To make a cutout, tape the cardboard to the flange. Trace the holes with a pencil from the flange to the cardboard. Drill out or cut out the holes in the cardboard. Mark the holes at the mounting point's surface using the flange itself or the cardboard cutout. Then drill out the pilot holes. I had to find shorter lag screws to for my temporary mounting in a 2-by-6.

2. After the holes are drilled, secure the mounting flange. Start all of the lag screws into the holes before tightening any one of them completely. Bring all the screws snug, and then tighten all of them in an *X* pattern and repeat once, being careful not to over-tighten them.

3. The mounting flange by itself is not enough to hold the dish in place, so anchor the braces in a similar manner.

4. With the braces fastened to a second piece of 2-by-6, orient the mounting mast pipe relative to the ground as close as possible to 90 degrees. For this, use an assistant and a small level to check the vertical position of the pipe.

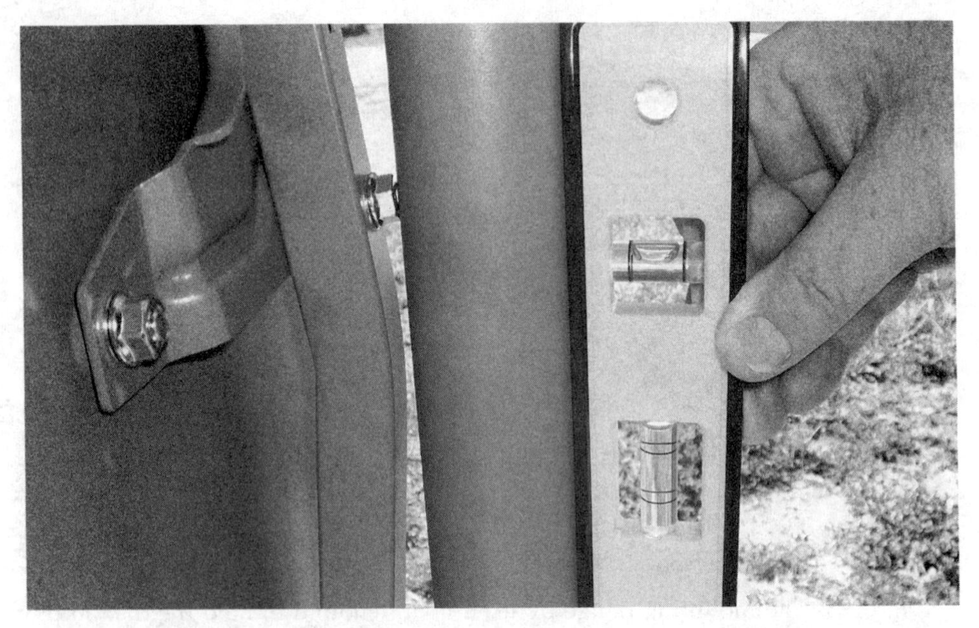

5. With the pipe at level and a helper holding it in place, drill an anchor for the braces to maintain the relationship at 90 degrees.

NOTE *This will be slightly different if you are mounting to a wall, for example. The braces can be placed against the wall surface. It is best to mount the braces directly into one of the wall's uprights, so use a stud finder to find the stud and long enough lag bolts to reach it. It can be a good idea to securely install a nailer board to the wall and attach the braces to that. Use lag bolts to secure the nailer board in the building's vertical studs. Nails simply will not hold a dish subjected to strong winds for very long, if at all.*

CAUTION *If bolting to a wall surface, be very careful to avoid electrical wires.*

Pole or Wall Mount

You could easily imagine this as a pole mount or wall mount instead. The process is the same. First secure the bracket to the pole, and then add the mounting braces. You don't need to bring the whole assembly up a pole or ladder, and that's not the way to do high mountings. Its weight is too much to handle and wind is not your friend when doing a job like this.

After you set up the assembly for mounting and know what pieces will go where, then remove the mounting bracket, take that up the ladder, and securely mount the bracket to its permanent location first. Then get help and two ladders to bring up the dish to finish the assembly.

You can also mount the mast pipe alone and braces next, and then bring the dish to place on the fully mounted flange, mounting mast, and braces you have already assembled. If working alone and without a bucket truck, this latter method would be preferred. Remember, safety first and use all necessary safety equipment. Use good old common sense in your approach and maintain safety at all times.

TIP *When doing this for the first time, assemble the entire dish assembly first. Then disassemble it into smaller manageable parts for mounting as previously mentioned. If you do more than one dish installation, you can develop your own procedures to get the job done efficiently as you gain experience.*

6. With the dish fully assembled on a perfectly vertical mounting pipe and mounted in its final location, you can completely tighten every nut, bolt, and screw except for the three that set the elevation and the two that will finalize the azimuth setting beginning with the butterfly flange.

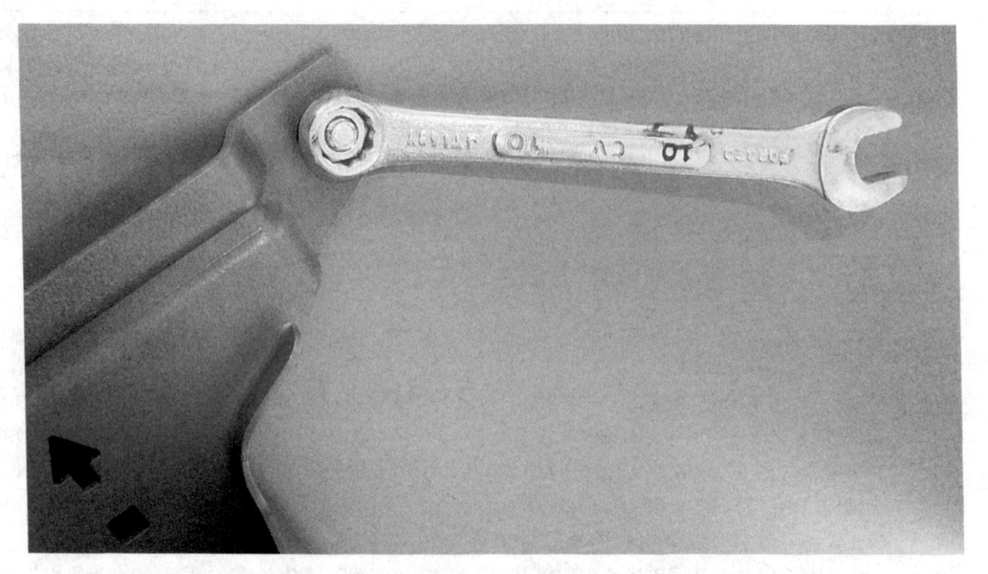

Are we there yet? Well almost! There will be more on the alignment topic in Chapter 7. There are just a few remaining tasks.

7. Use the compass to align the azimuth and look angle (elevation) for the particular satellite in which you're interested. After that is done, and the two pipe clamp bolts are fully tightened, you can set the elevation for that satellite. The lead edge of the pipe clamp that can be seen through the arc

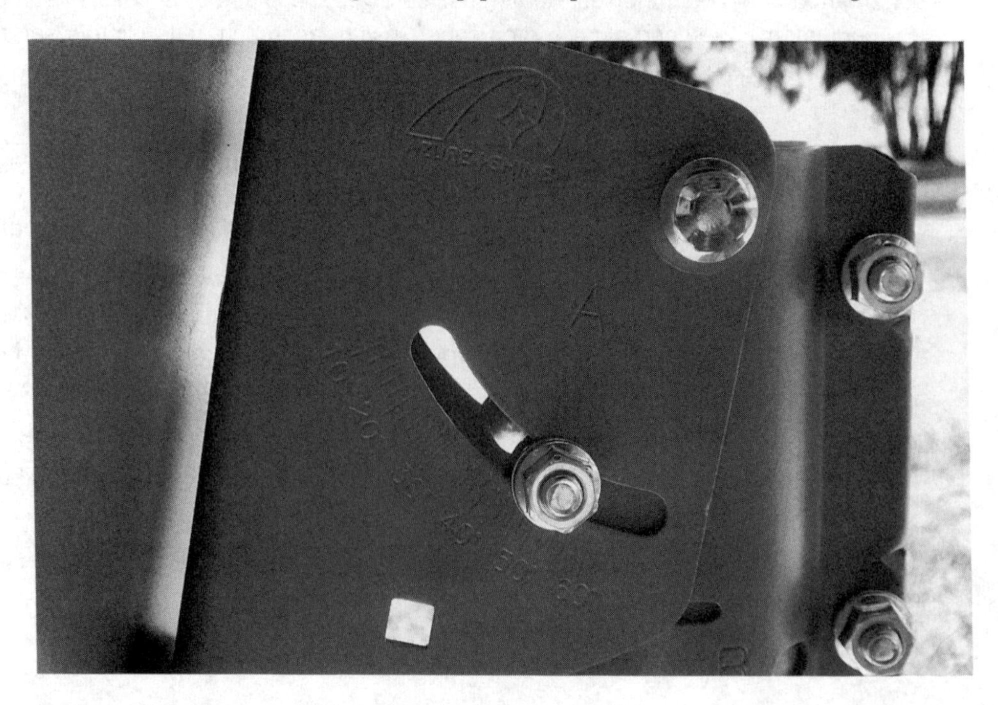

slot with the degrees marked sets is the angle pointer. It's not a bad idea to brush a drop of paint on the leading edge to help identify it. In the previous illustration, you can see that the elevation is 42 degrees. (If you purchase a motor to direct the azimuth and look angle, you will not be doing this step.)

8. Tighten all three serrated flange nuts to maintain the elevation setting. Do a final check and tighten all of the nuts, bolts, and mounting screws over the entire assembly.

9. The LNB distance to the nearest surface of the front of the dish is somewhat adjustable from front to back inside the plastic mounting bracket. It might need some tweaking from front to back later to improve reception. For now, align it so that it is equidistant, centered on the adjustment area, and completely vertical with the wire connection at the bottom, and snug up the Phillips-head screw going through the clamp just tight enough so it cannot move.

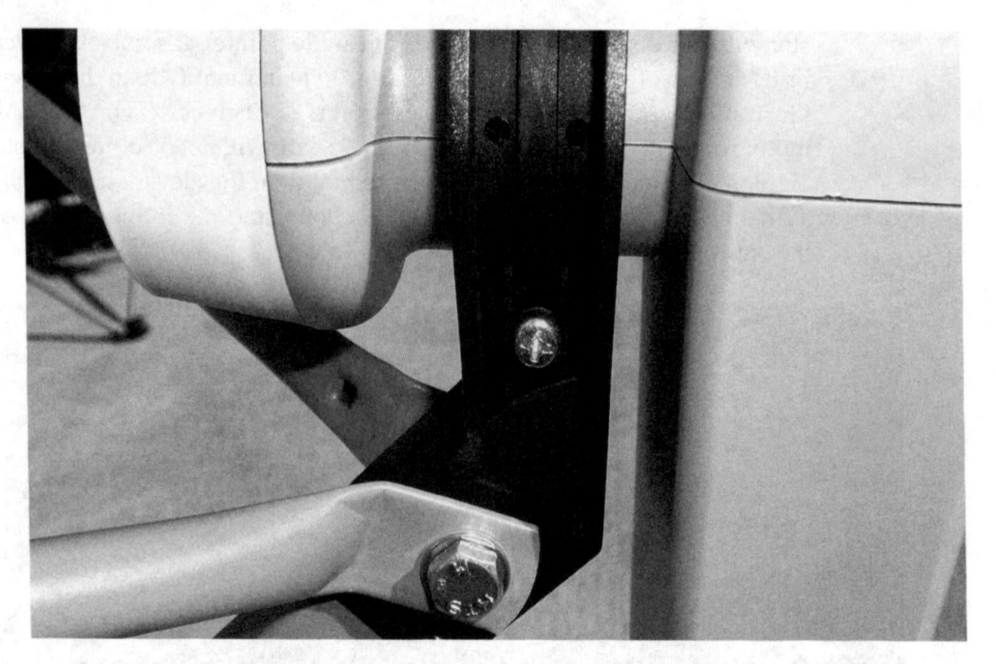

Take a few steps back, admire your assembly handiwork, take a picture for the scrapbook, and get a latte and a custard tart. Have a seat and enjoy. It's high time for a break.

CHAPTER 5

Necessary Accessories

When my childhood home first had telephone service, we would pick up a receiver handset and an operator 3 miles away would see an indicator light up on a matrix of jacks at a switchboard. The human operator, Gladys, would plug her headset into the jack below our light and say, "Number please?" We'd give Gladys the number (or even the person's name), and she would plug in our line into the jack associated with the number we needed, while applying the ring voltage to the appropriate line pair.

If the recipient of your call was home and answered the phone, you would be "connected" for the duration of the call. In those days, it was difficult to not get caught calling the corner grocery store to ask whether they had "Prince Albert in a can," only to tell them to let Prince Albert *out* of the can. That fun prank did not come with total anonymity until the rotary dials took over the phone system.

You might believe that hooking up accessories and peripheral satellite receivers, VCRs, DVDs, and sound systems to a TV is almost as intimidating as being thrust into the role of a 1950s' era telephone operator's job. But after reading this chapter, you should feel confident enough to connect any number of devices to a television receiver and perhaps be willing to be the "Gladys" who connects up someone else's system.

This chapter will help you work through the details of connecting electronic devices to become a functioning part of your total TV viewing system. The step-by-step instructions and concepts will help to eliminate any confusion about what type of cable or connection should be used and how it should be connected. The following sections provide the contrivances required for routing this signal over various cables and through other devices on its way to your television screen.

Airwave Inputs

The airwaves provide the media in which the video and sound signals travel to reach an antenna or a dish. The term "airwaves" is used, but the electromagnetic signals can travel through a vacuum and the emptiness of outer space. The signals travel though the atmosphere even though the atmosphere provides resistance to the signals, such as from thick cloud cover or electrically excited gases. Elements in the atmosphere also contribute interference by way of lightning flashes. To receive the best signal possible from the airwaves, you need to have a decent antenna properly secured and properly aimed to maintain the highest quality of signal.

The cable ANT input jack on the television shown in Figure 5-1 (upper-right corner) is the source connection for transferring the electromagnetic signals from the antenna OTA digital signals, cable television signals, or for analog signals from video devices that produce analog outputs typically on analog channel 3 or 4. The cable to use for this was traditionally an RG-59/U coaxial, and it is certainly still OK to use that on any device using these lower frequency ranges. If you are buying new cables or making your own to length, it is easier to buy and use RG-6 (one type of cable), because it will work for analog, digital OTA, and satellite and with one or two types of F-connectors for terminations: either straight and 90-degree offset con-

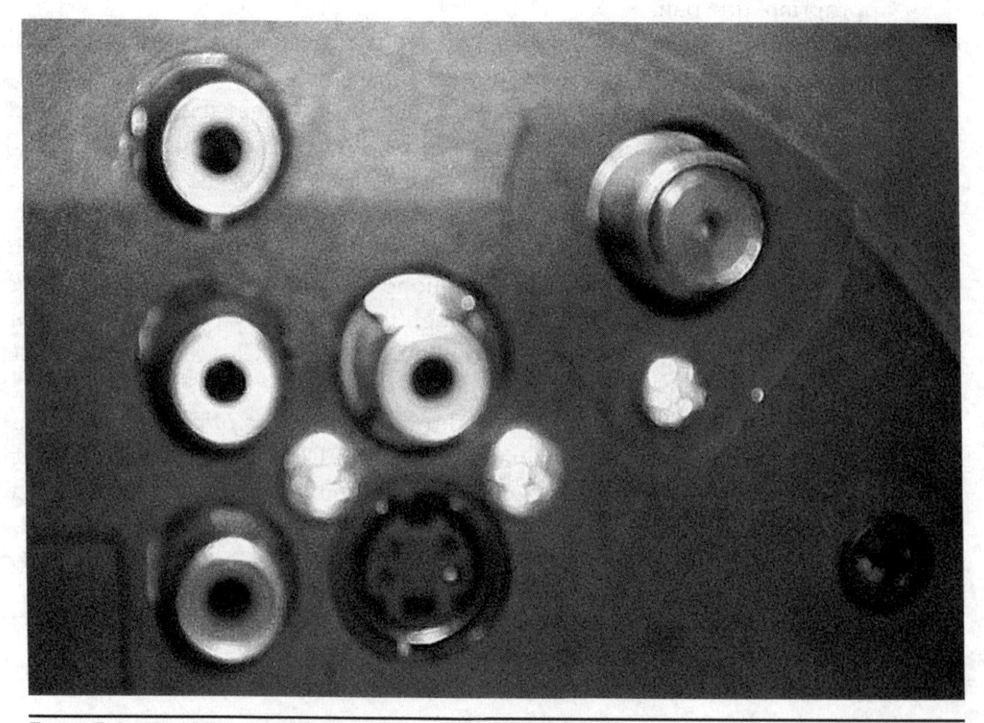

FIGURE 5-1 TV ANT threaded connection for F-connector

FIGURE 5-2 ANT in and loop out on satellite receiver

nectors. Use the 90-degree offsets where space is limited, such as on the back of equipment in entertainment center cabinetry.

For satellite down-leads, always use RG-6 and sealed F-connectors. Figure 5-2 shows the two cable connections on the S10 receiver with the connection for the dish/low noise block (LNB) already attached with a cable. The *loop out* (the empty slot in the figure) passes the signal to a downstream receiver if there is one, providing a means for two receivers to share concurrent use or alternating control of the same dish.

Device Outputs and Cable Types

Getting the desired video signals to your TV screen and sounds to your speakers or headphones with the best possible quality is made possible by hooking up the available outputs to the best option inputs for viewing. A secondary objective is to be able to route the video and audio to a recording device for viewing at a later time. Sometimes you will want to view and record at the same time.

The last analog recording format was the VCR. Most recordings are now stored in a digital format, either on DVD, Blu-ray, CD, jump drives, microchips, or hard drives. Whatever the recording media, the data file with the video and audio information must be processed through a decoder device to produce a video signal and associated audio signal that can be further processed into on-screen video and audio by the TV set (or sound processor). Regardless of the playback device, the output connections will typically use RCA jacks (sometimes called sockets) for making the data output link to the TV. There are two other possibilities: older VCRs, which output analog signals on channel 3 or 4 and use cable connections; and a subset to a recording device's removable storage media such as portable hard drives or jump drives that will most often use USB connectors.

Figure 5-3 shows the back of a currently available model of a video playback device. This is a good example of the need for RCA plug-end cables. Along with the high-definition multimedia interface (HDMI) output jack are all of the commonly used RCA output jacks. A color coding is associated with them that you can't see in the black and white photo. On the top row the (composite) video out (first on the upper left) is yellow. Next in the top line are the component video output jacks Y, Pb, and Pr, color coded green, blue, and red, respectively. The bottom row output jacks R (analog) and L for stereo output are red and white, respectively. The coaxial digital is orange and is used for connection of multichannel sound (5.1) to a sound processor.

FTA receivers will typically have a number of outputs as well, providing choices for connections to a TV's inputs.

Figure 5-4 shows a high quality pre-molded cable with RCA jacks color coded for RBG (red, blue, and green), obviously intended for use with component video connections. The color coding just helps you match the outputs and input plugs to the correct jacks on the TV and playback device.

The USB cable jack in Figure 5-5 was developed to support serial communications for computer to communicate with peripheral devices such as printers, external drives, modems, cameras, and jump drives. The USB connection has found its way into the video/TV/FTA and cable TV world because of digital video recorders (DVRs) or interface boxes and receivers with DVR capability. It seems that no matter how large internal device storage is, it is never big enough, and of course a subset of the population is willing or wanting to share stored video files. The USB standard and cable connection solves both issues quite nicely. Portable hard drives and the smaller jump drives can be easily used for adding storage space on the FTA digital video recorder features by working through a few setup choices on the

FIGURE 5-3 RCA output jacks

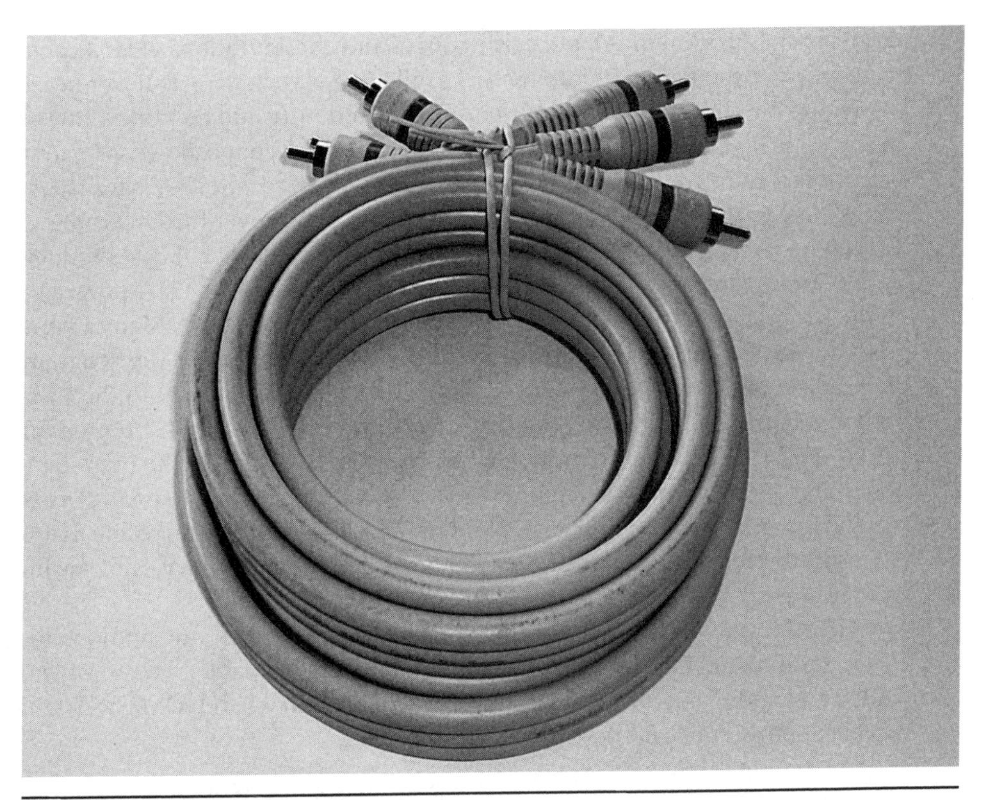

Figure 5-4 RCA jacks/video cables

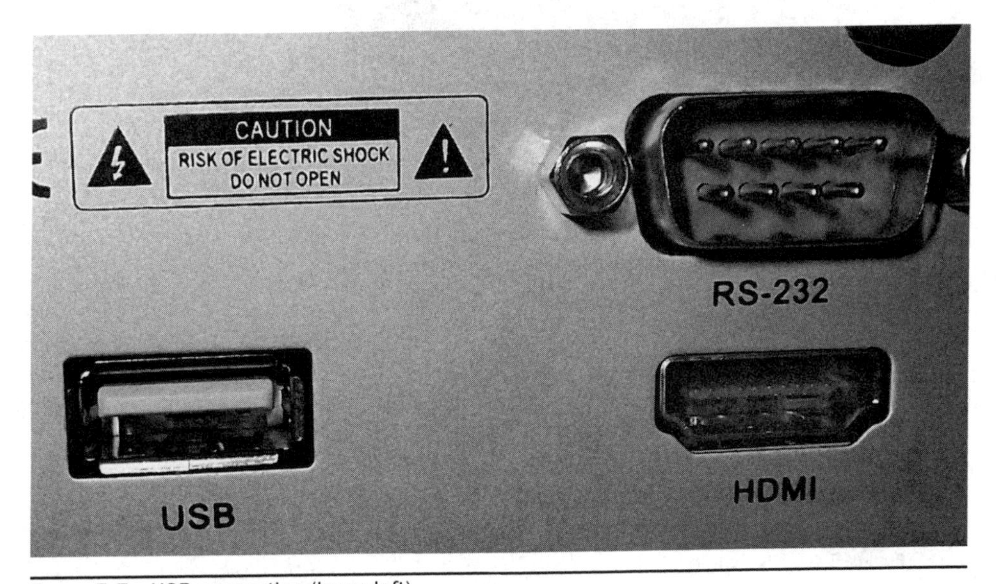

Figure 5-5 USB connection (lower left)

receiver's menu system. All the various sizes and configurations of USB cable connections are designed to fit into the jacks only one way; forcing is never necessary.

Figure 5-6 shows the USB cable end that would plug into the connection shown in Figure 5-5. A 500-GB portable hard drive with a USB connection can be found on sale for less than $60, making this an excellent way to record shows for later viewing. Power to the recording device and data both travel over the USB cable.

Internet-ready devices, which today comprise a whole gambit of products that center around home television entertainment from FTA and OTA programming, gaming, and recording, will require a connection to the Internet. Many FTA receivers also use the Internet to capture and download upgrades to the programming features or core operating system that runs the receiver. Check with the FTA manual or manufacturer to learn whether Internet OS upgrades are supported on your receiver. The alternative receiver upgrade path is to download the upgrade with a computer and transfer the upgrade via a traditional RS-232 serial port on the receiver, also shown in Figure 5-5. Figure 5-7 is a CAT 5 computer cable that would be used to connect FTA receivers or other Internet-ready devices to an Internet router or modem.

HDMI is now the preferred connection for receivers, TVs, and audio video payback equipment. Figure 5-8 shows a standard HDMI cable. These cables carry video and audio and are the best choice for connection when HDMI jacks are available on both source and destination equipment.

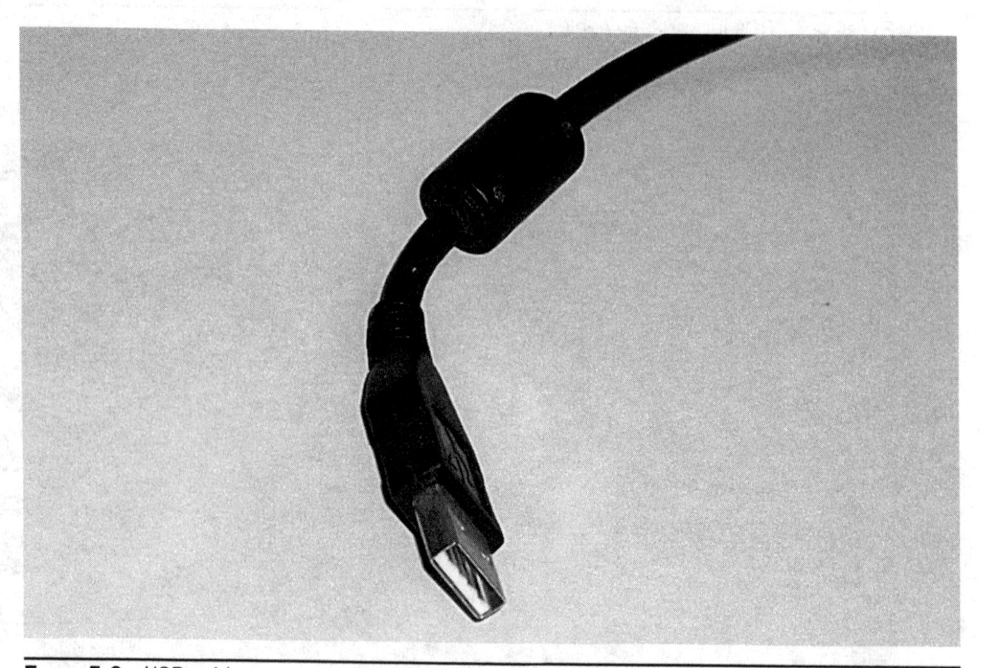

FIGURE 5-6 USB cable

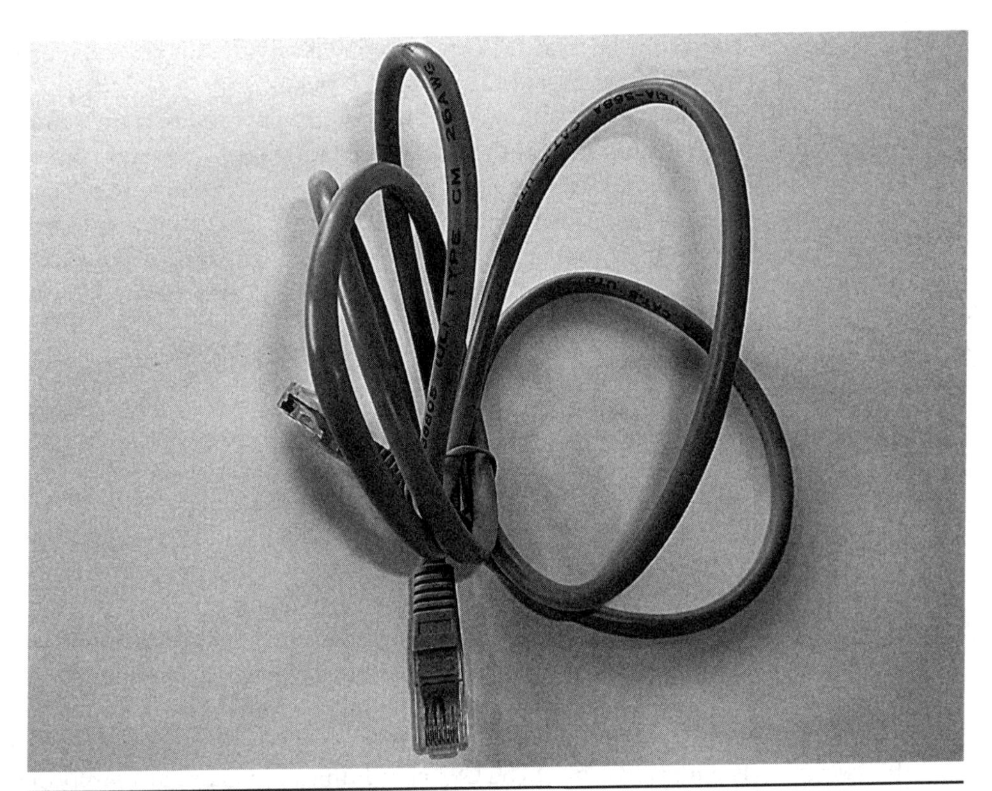

Figure 5-7 CAT 5 network cable

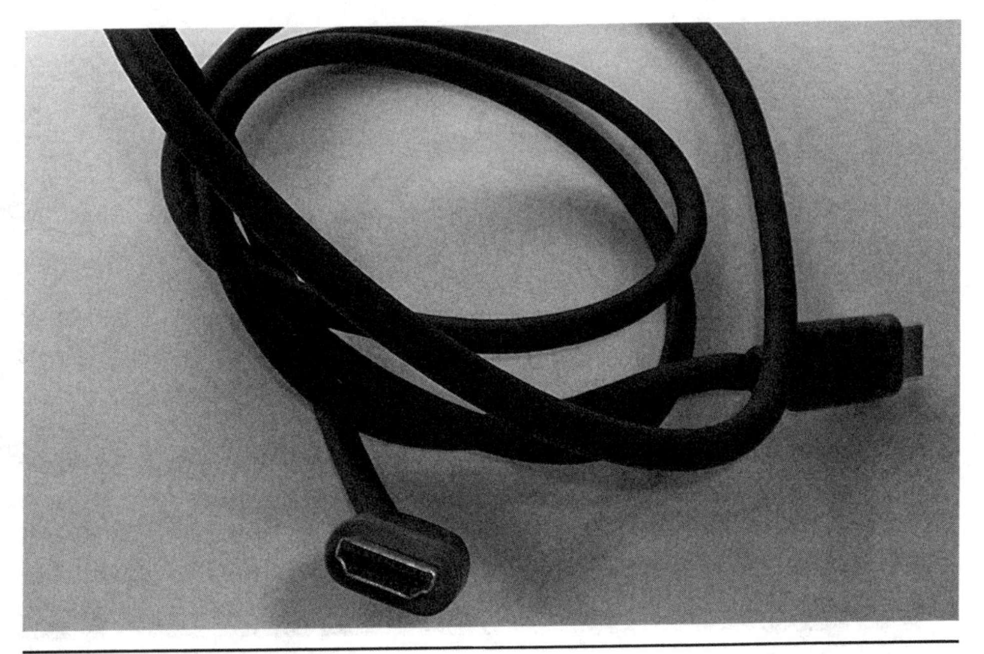

Figure 5-8 HDMI cable

Figure 5-9 shows the back of the OpenBox S10 receiver with all of the connections populated with cables except for the RS-232 jack.

The most common cable set in use on video equipment today is the composite cables. Old and new equipment both frequently feature this option for connection. As with any new communications standard, HDMI will take time to become the dominant connection. In the meantime, you might need to have a few of these composite cables shown in Figure 5-10. They are color coded, yellow for video, and white and red for stereo sound, to connect your recording and playback devices to your TV viewing system.

Cable Quality

As a full-featured video experience system is brought together with FTA, OTA, Internet, and recordings of all types, a few different kinds and styles of cables will be needed to make those connections. The big box store salespeople might try to convince you to "buy up" when it comes to cables, in hopes of selling you the more expensive gold-plated varieties. They are nicer, no question, if not prettier, with shiny gold plating, but it is hard if not impossible to discern any difference in performance over the short runs usually needed at the home entertainment center, often with everything being in the same cabinet or on the same wall. The one notable exception is with USB cables. The USB communications standard protocol is not

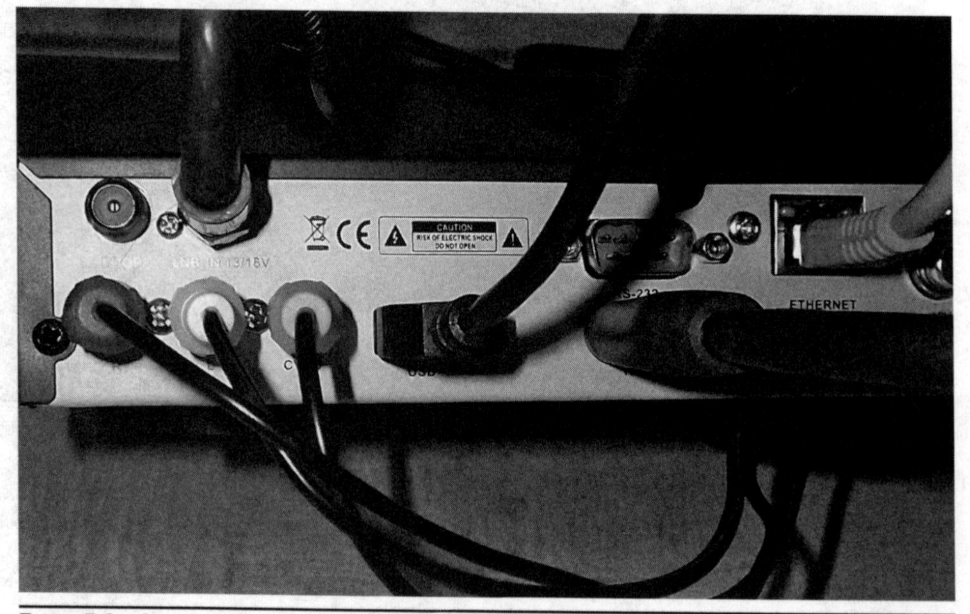

FIGURE 5-9 Connections used on the OpenBox S10 receiver

FIGURE 5-10 Composite video cables

very tolerant of poor quality cables or cables that are longer. For all other cables, the less expensive "run-of-the-mill" types are fine for these short runs. If you are buying on a budget, thrift stores are often gold mines for finding both everyday and higher priced cables for pennies on the dollar. (In addition, your thrift store purchases provide cash for charitable causes and keep serviceable items out of the landfills.)

Making your own cables to standard with rated materials is also an alternative to paying more than necessary on the way to a quality outcome. The section that follows details making your own cables for down-leads and jumpers with RG-6 and straight F-connectors. Universal compression crimping tools or those with interchangeable dies will typically work with RG-6, RG-59, RG-58, BNC, and RCA fittings from a number of manufacturers, making it possible for you to custom size and build connecting satellite, OTA, and video/audio cables.

Assembling RG-6 TV/Satellite Cable with Seal F-Connectors

The most important cables for FTA satellite reception are the down-leads and interconnections on the FTA satellite dishes and receivers along with any interconnections between LNBs to switches or in-line amplifiers. You can buy these cables or pay someone else to make them up to length from bulk for you. The easier and least costly way is to make them yourself on site, to the exact lengths needed for the individual installation.

Three tools are needed to cut the cable to length and assemble the F-connectors on each end: diagonal cutting pliers (or a coaxial cable cutter) to cut the cable to length, a cable stripper to prep the ends, and a crimper to clamp the F-connector to the cable end.

Begin the process by becoming familiar with the cable stripper. The one shown here operates with two blades under spring tension that can be seen at different depths in the hole for the cable. The top of the stripper acts as a button to depress the blades to make the hole big enough to insert the cable end. The cable is inserted in the end shown that is closest to the lower set blade.

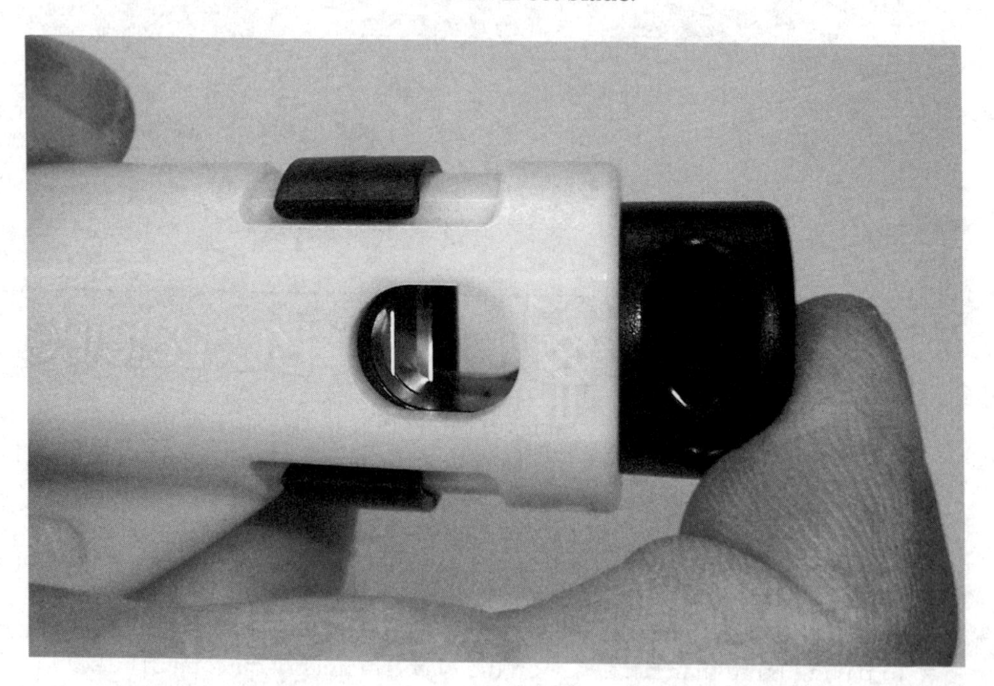

1. Carefully insert the cable into the hole in the stripper while pressing the top button to depress the cutting blades.

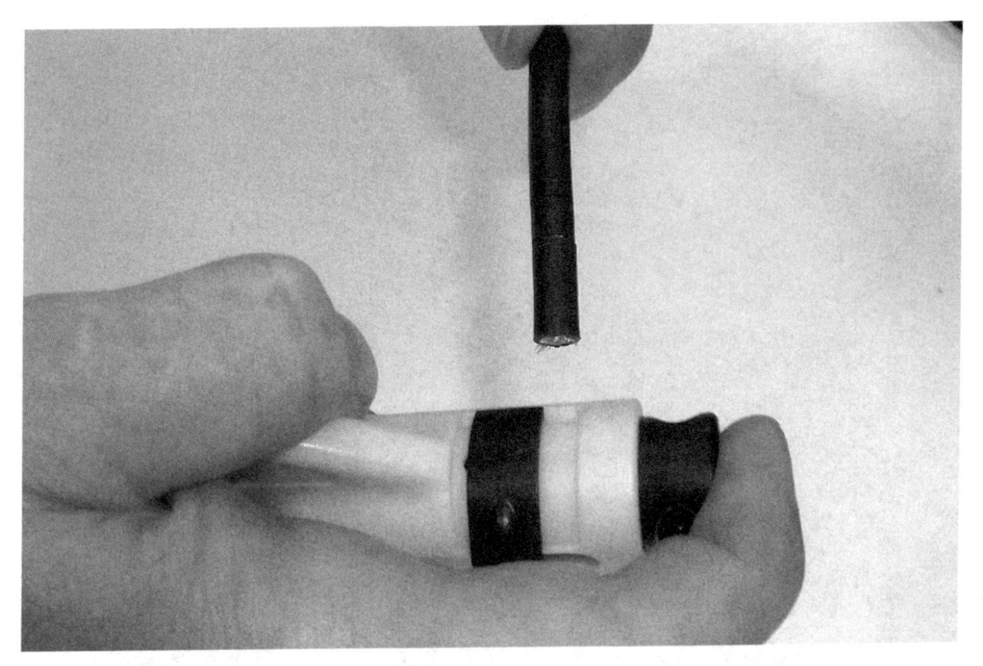

2. While holding down the button, push the cable through the stripper so that it protrudes beyond the stripper by at least 1/4 inch.

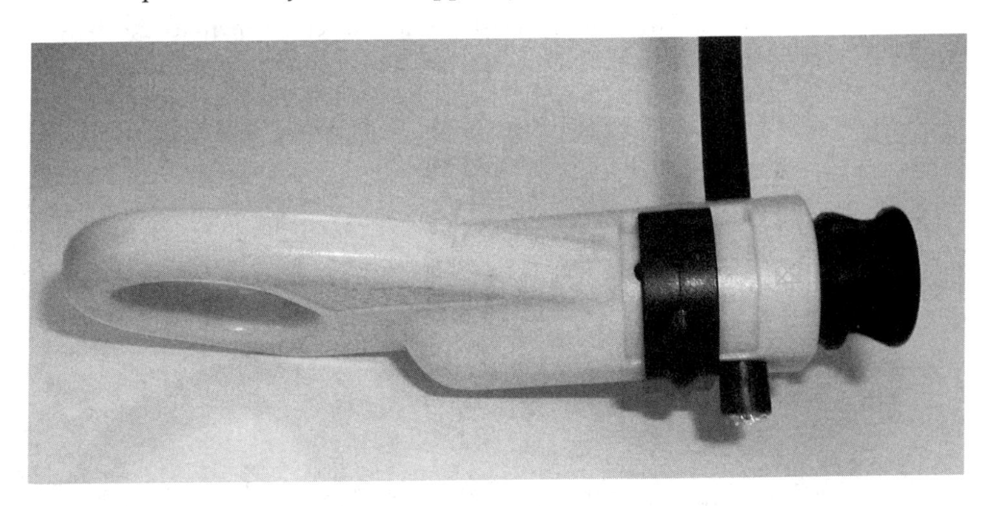

3. Release the button and rotate the stripper more than a full 360-degree circle, as shown next. The two cutting blades will make their precision cuts through the sheath, shielding, and insulation.

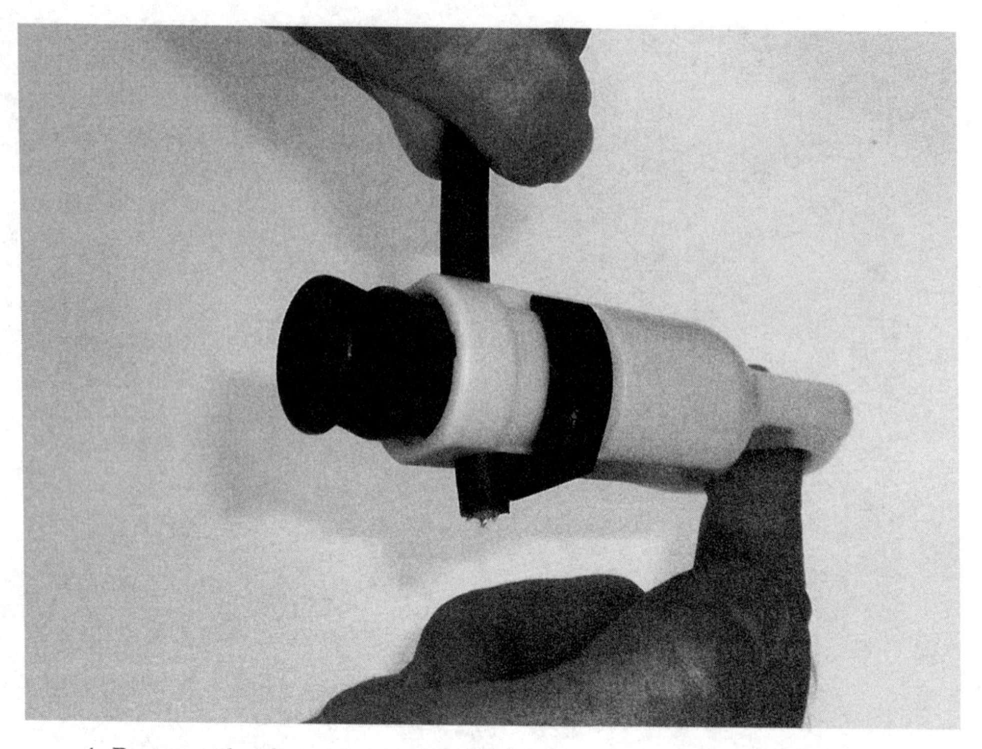

4. Remove the first bit of waste from the end down to the bare center conductor.

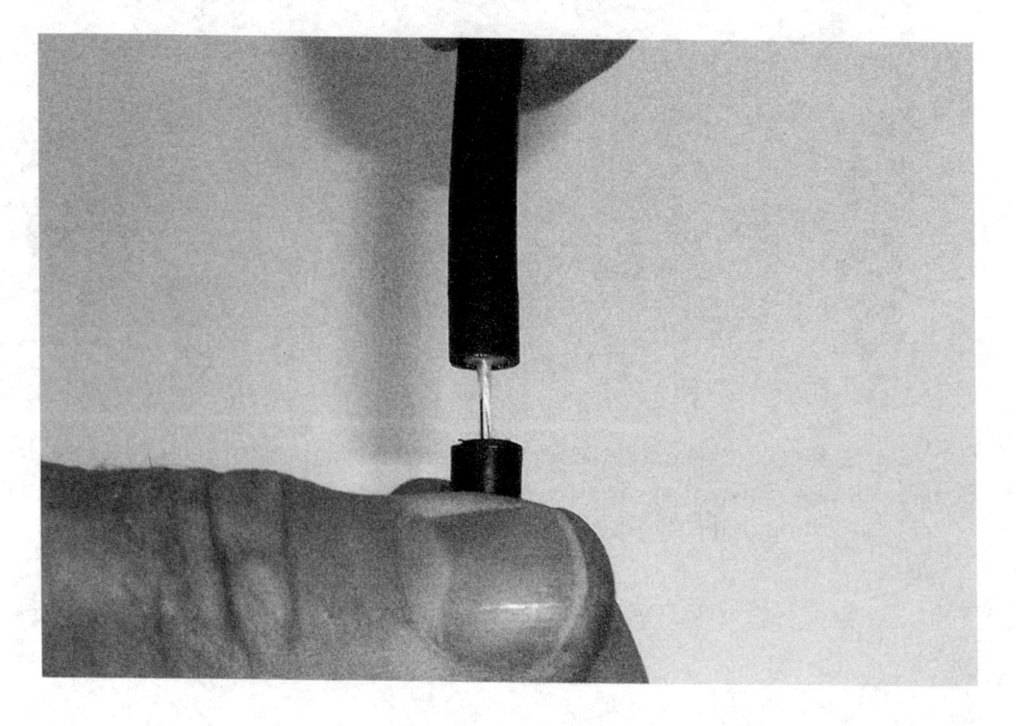

5. Remove the second bit of the wire's outer sheath, exposing the outer shielding.

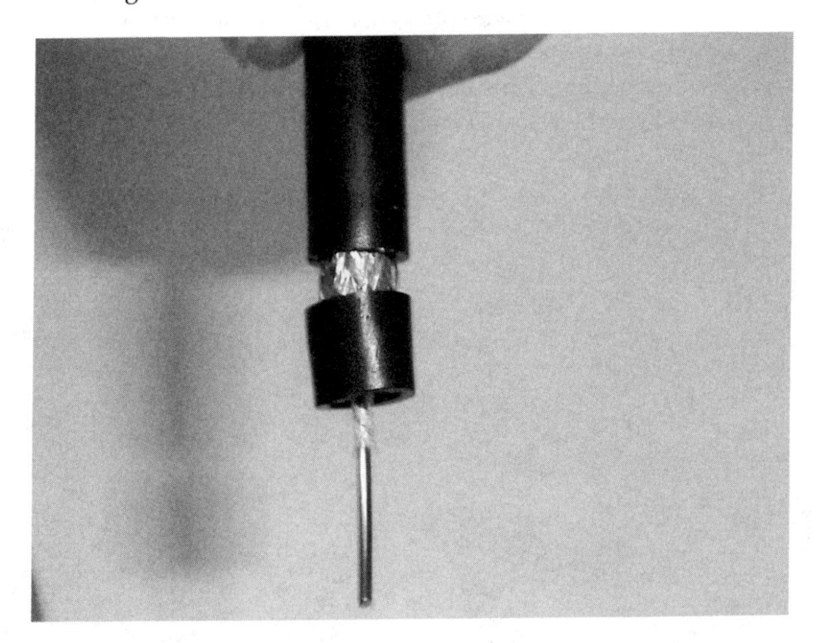

The next illustration shows the exposed center conductor with some of inner wrap still on the conductor, which must be removed along with any loose strands of the shielding. The shielding strands are small and hard to see, so removing them now to prevent a short condition later is critical.

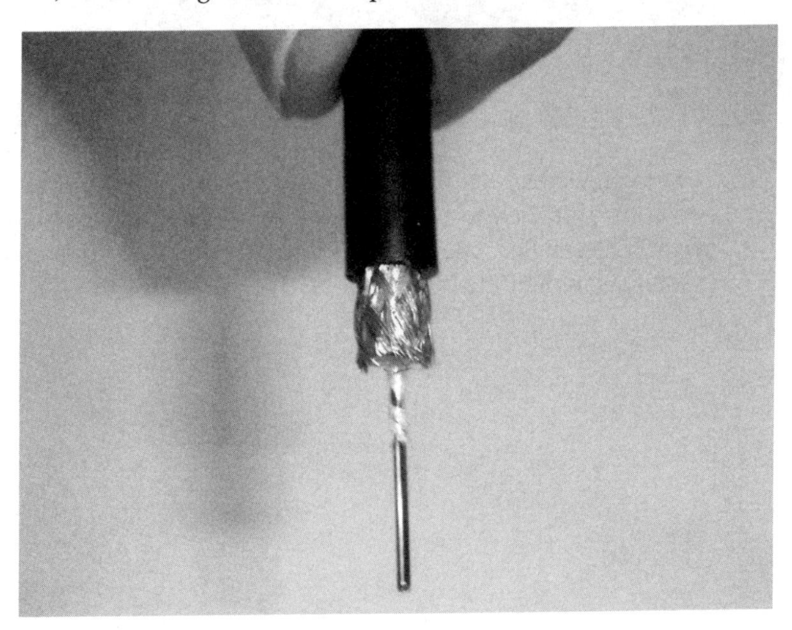

6. Fold back the shielding and check the center conductor for correct length. The dimensions for proper preparation of the wire for the F-connector require cutting the wire off so that 1/4 inch (or just over 6 mm) of the wire protrudes beyond the inner insulation material. The length of the remaining exposed inner insulation should also be 1/4 inch, making the folded-back shielding also 1/4 inch long.

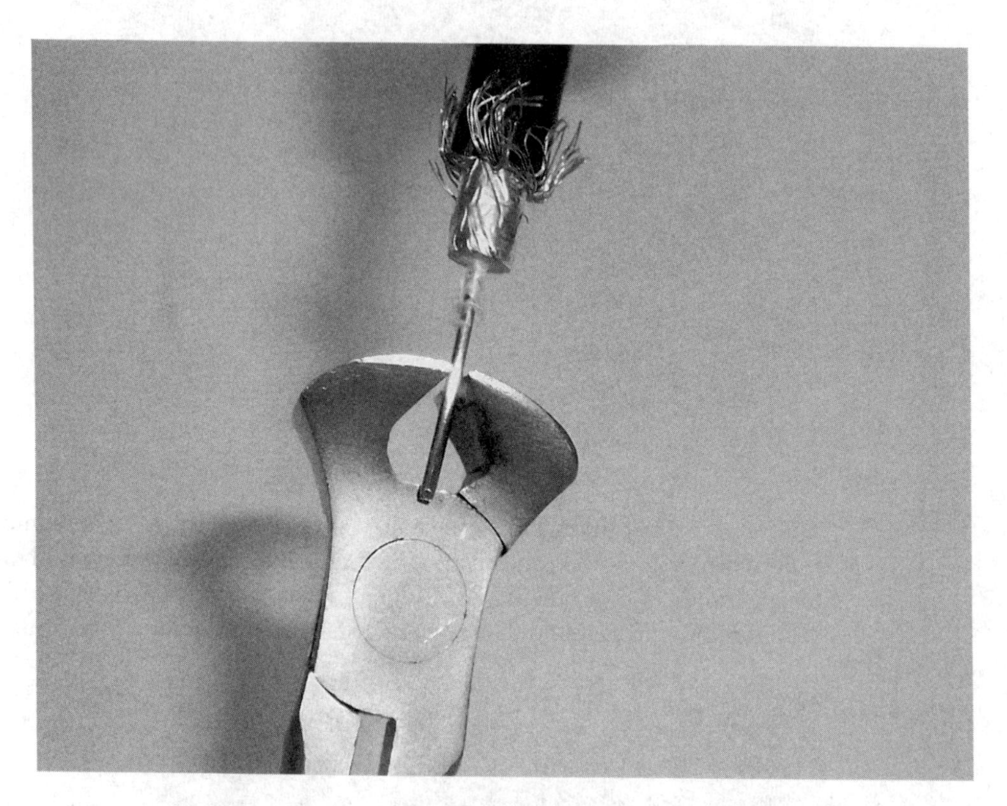

The connection end of the F-connector is machined to two barrels, the larger one threaded to screw connect to the equipment jack or splice connector. The smaller inner barrel is just big enough in diameter to hold the inner insulation of the RG-6 cable.

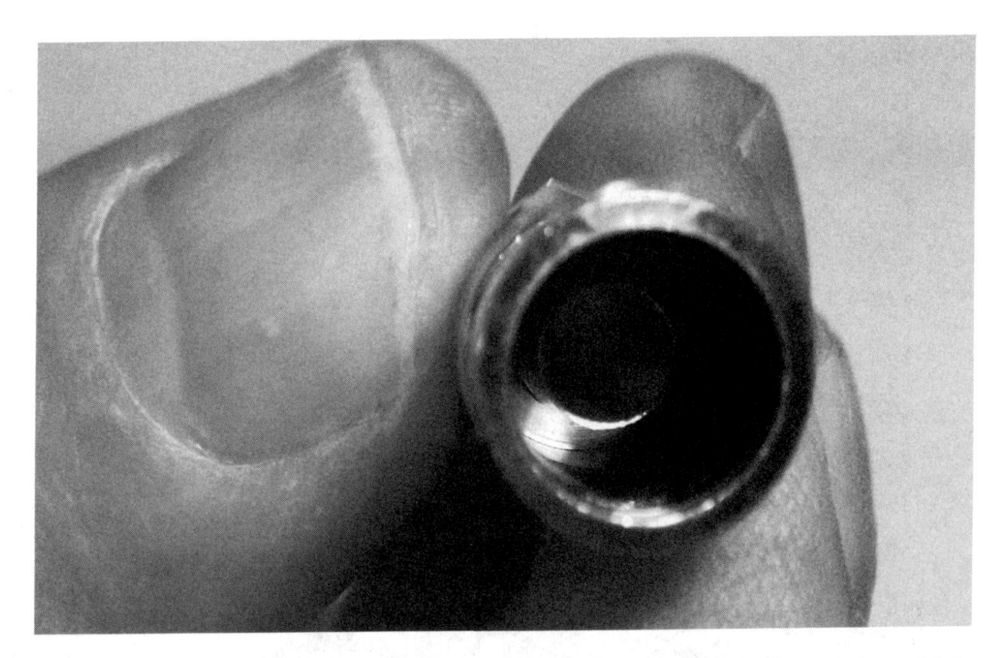

7. Insert the cable into the F-connector until the leading edge of the cable's inner insulation is even with where the small barrel (front) ends.

8. Place the cable and F-connector into the jaws of the crimper, centering the F-connector on the adjustable F-connector die shown at the bottom of the next illustration. The slotted prong at the opposite end of the crimper will apply pressure on the back end of the F-connector to insert it and its seal into the barrel.

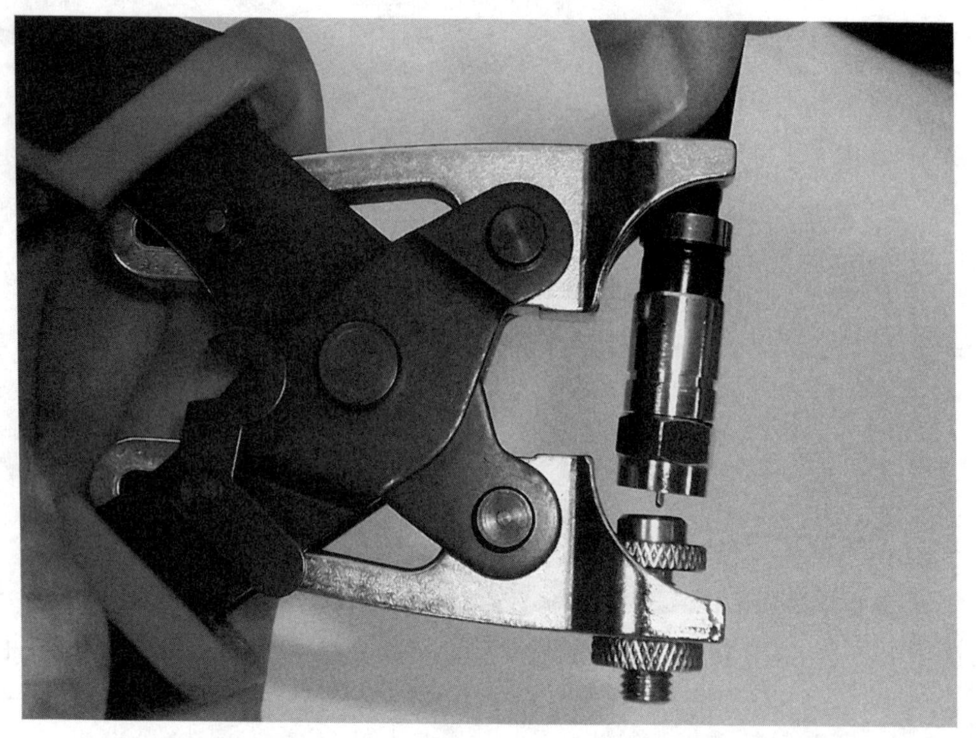

9. Slowly apply pressure with the handles of the crimper, being sure to maintain the alignment of the die, the F-connector, and the slotted prongs of the crimper.

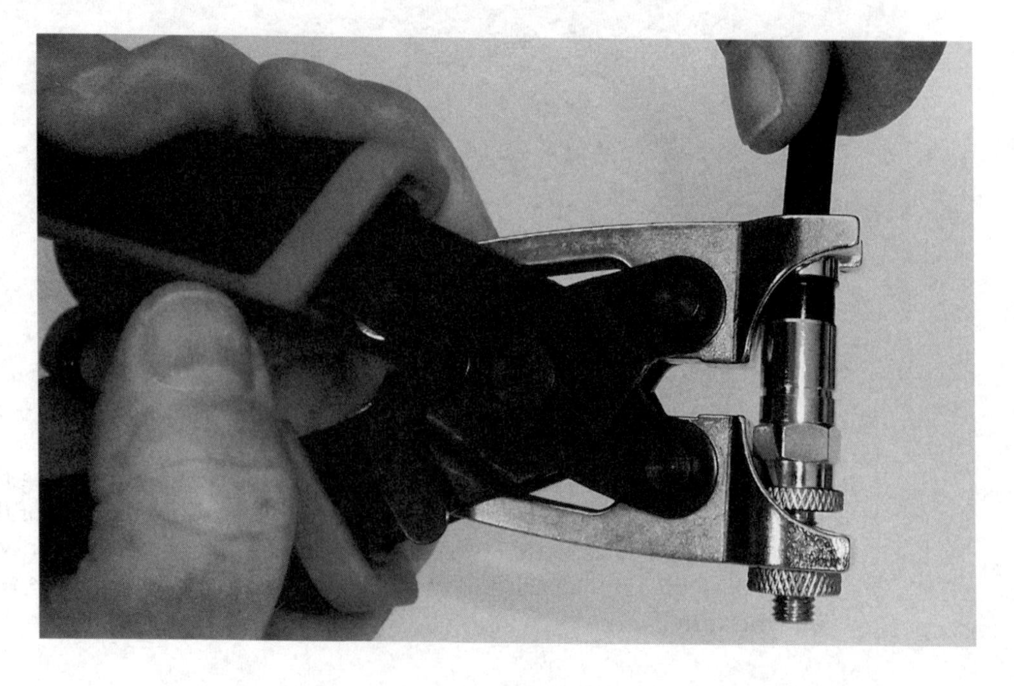

10. Continue to squeeze the handles of the crimper, pressing hard until the sleeve and seal material on the back of the F-connector is fully inserted into the back barrel of the F-connector.

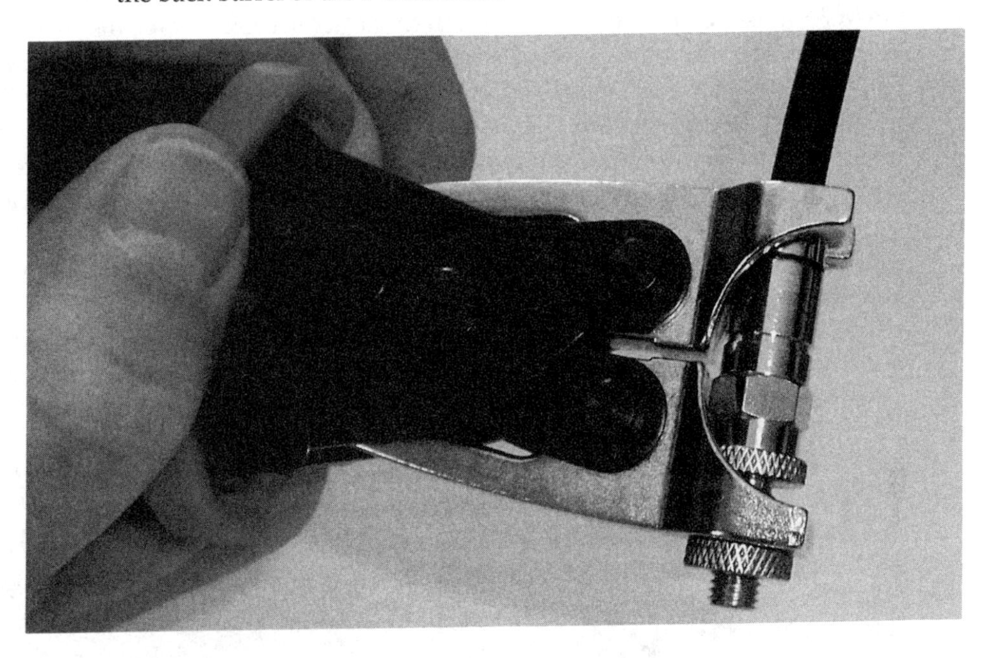

11. Remove the connector and cable from the crimper and inspect to make sure the center wire was not bent and that the connector is fully crimped together. The final product should look much like the next illustration, ready to connect to a jack on the receiver, LNB, motor, or on a switch.

After you've made the cables, they will need to be secured about every 3–10 feet along the runs with cable clamps, staples, or straps, depending on what the run crosses. In some cases with overhead runs, a messenger ground wire integrated in the RG-6 must be used to provide a path to the service ground for the mounting mast pipe. The messenger ground wire for grounding the mast is integrated into that specialized type of cable. Use the grounding messenger wire cable type when independent ground wire runs are not practical. On overhead runs of any length, it might be necessary to use a steel suspension wire to keep wind and weight stresses off of the connections. The suspension wire must be big enough to do its job to carry the weight of the signal cable and integrated ground wire.

The next illustration shows four different kinds of staples for fastening wires and cables to wood surfaces, studs, or joists. The bottom two in the picture are single nail and are intended for holding shielded cable. In tight spaces, simply pull out the nails and use small screws instead. Replacing the nails with small sheet metal screws will allow you to attach the cable and staple to thin metal surfaces.

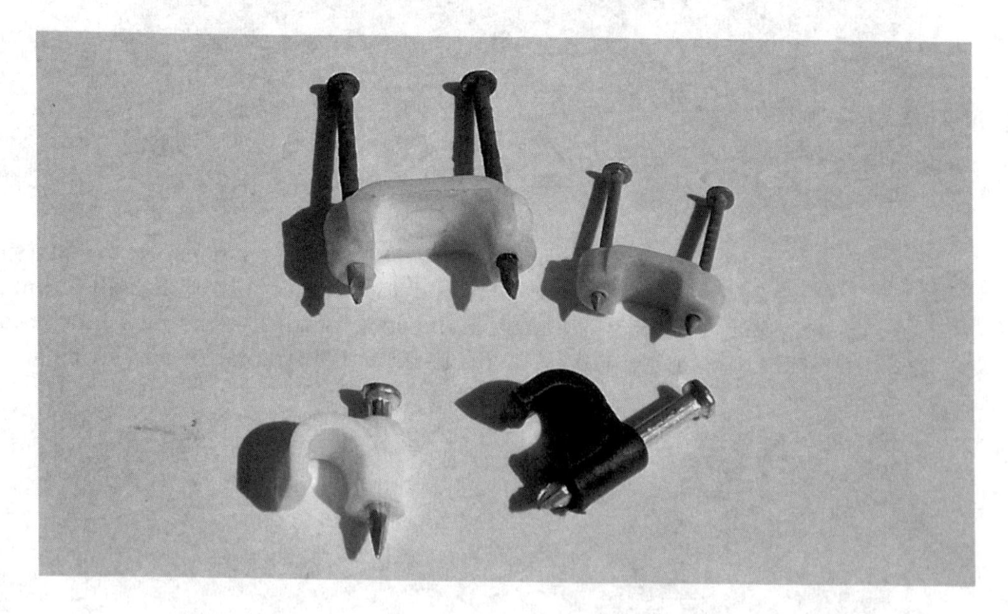

Another product for holding cables and leads in place is the cable tie, shown next. Just wrap it around the cable and a pipe, eyelet, or other object, and insert the end (as shown at bottom) into the clasp and pull tight the loose end. The notched surface on the tie must connect to the cog in the clasp.

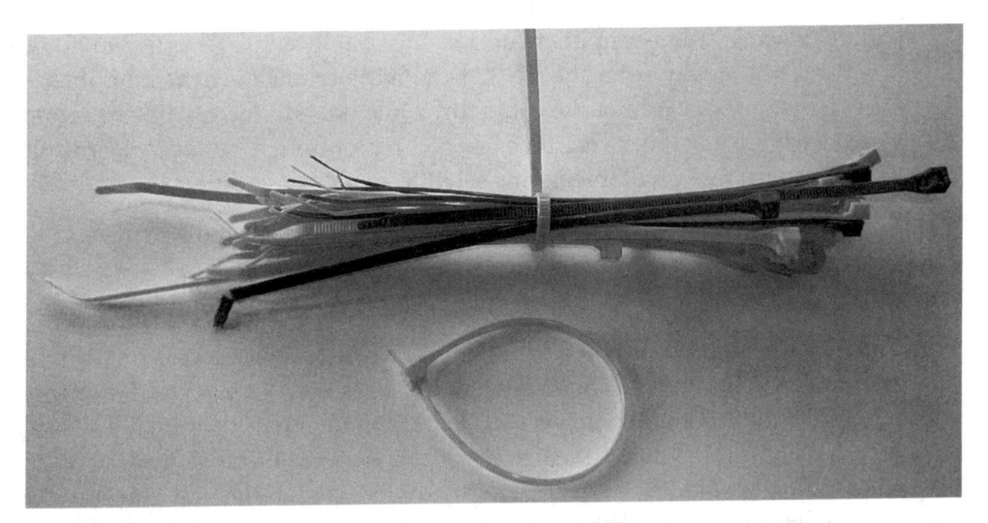

Making the cable terminations yourself takes only a few minutes if you use the right tools and will provide the exact lengths necessary for your installation. Be forewarned, however, that having the tools and the knowledge to do the cable terminations might make you the go-to person among the TV programming devotees in your social network.

Switches

Typically, only two types of switches will be considered for use on satellite cable runs. Refer to Chapter 8 for details. Avoid use of A-B switches on satellite cable connections. These are OK for switching unamplified OTA or cable signals if necessary.

Amplifiers

Long down-leads are occasionally necessary for getting the FTA signals to the receiver inside a building. You can get the best signal when long cable runs are necessary in three ways:

- Use a very efficient LNB with the most gain (sensitivity).
- Use the larger size dishes (75, 90, 120 cm) to collect the maximum signal strength.
- Use an in-line satellite intermediate frequency (IF) amplifier placed between the receiver and the dish LNB.

The in-line amplifiers get their power from the receiver to run and pass the tone signals and power for LNBs and motors through with a feature called *DC power passing*. Placing in-line amplifiers on a single dish to single receiver would be best at a third of the way from the LNB to the receiver, pushing the amplified signal over the longer distance of the cable run while not adding any line or connection losses early on for the signal tones or DC pass through voltages. Digiwave and EMP-Centauri make suitable in-line satellite signal IF amplifiers.

Junctions

Cable junctions are common with cable TV distribution and occasionally with distributing signal from OTA antennas. Junctions are not used with FTA satellite signal distribution: a multiswitch is used instead with individual down-leads for each receiver. Figure 5-11 shows the before picture of a very nicely done double junction for a cable installation that brings in the cable signal to the first junction block on the left, which correctly splits the cable off right away with one line to the IP phone modem, another to the Internet cable-modem, and the third to a second junction block on the right to distribute cable TV signal to four TV locations in the house. All runs in this case were nicely done with RG-6 cable.

Figure 5-11 Cable distribution box with junction blocks

This situation provided the perfect opportunity for this homeowner to disconnect the cable ties to the TVs from the junction block and to use this location and box to install a 4-port multiswitch. All that was necessary was to bring the two LNB down-leads from the ground-mounted dish into this box connecting to the LNB ports on the multiswitch. Then the to-TV leads from the junction block were moved over to the Rx1 to Rx4 ports on the multiswitch. If the next homeowner wanted to go back to paid cable, everything is there in one place to use over for the FTA installation, or to switch back to the cable junction box once the FTA receivers were removed.

If you already have cable and are switching to free FTA, check to see if your situation is similar to this one, which allows you to leverage much of what is already there for cable runs. If your home's existing cable runs were done with RG-59 or older cables, replace them with RG-6 for optimum satellite signal transmission.

CAUTION *All junction blocks are not made equal. Look at the frequency range ratings on them as well as dB loss ratings. The signal in-out labels should be followed also. Check the ratings to ensure that the frequency ranges are appropriate to the signal range you need for the application. Use as few junctions as possible, making the star configuration the best (refer to Figure 5-10) as opposed to long serial runs with branches.*

RF Modulators

You might not see it used often, but an RF modulator might be useful for providing video to a set in your home system. These modulators typically have inputs for composite or S-video connection and stereo inputs for sound, and they "broadcast" that signal over analog channel 3 or 4. Models are available to transmit on higher numbered TV and cable channels as well. HD versions are available but too expensive for most home applications. Figure 5-12 shows an RF modulator that has a pass-through connection to the OTA TV antenna. The device uses an 110V AC cord for the processor's power. You might use this to send video/audio to a TV some distance away, or you might have an old playback device you want to use by connecting via an analog channel.

Getting the cables and connections right is simply a matter of paying attention to the details. You need to map out what goes where from the source information that shows what is available and match what is available on the TV or sound processor with what is provided as outputs from the recording and playback devices included in a full-featured home TV viewing space.

Once the connection needs are identified, you're ready to acquire and connect the few specialty cables necessary and sit back and enjoy the show.

Figure 5-12 RF video modulator with pass-through OTA connection

NOTE *The prototype selection for the demonstrations and images in this book is the OpenBox S10 (OpenBox S10 [Sclass] HD8899MAX DVB-S2 TV Receiver). This unit went into production in early April 2010 and is aimed at the North American market for FTA, but with the right power cord it can work anywhere in the world on 220-volt European power. Selection of this unit was based on its supporting most of the features and functions summarized in this chapter. For a complete listing of the specifications from the exporter of the OpenBox S10, see www.alibaba.com/product-gs/374424759/Openbox_S10_Sclass_HD8899MAX_DVB_S2.html. The URL is for volume buyers and not for one- or two-off consumer purchases.*

CHAPTER 6

Guide to Satellite Receiver Selection

You'll find significant price versus value tradeoffs when choosing from among all of the currently available models of FTA receivers. When you're selecting between an entry-level model receiver and a top-of-the-line unit, the highest prices can be double or triple those of the lower priced models, even those from the same manufacturer.

The biggest price increment decision for many users will be choosing between high definition TV (HDTV) or standard definition TV (SDTV) for viewing your favorite FTA channels. Few HD channels on FTA are available now, but the number of available HD broadcasts is likely to increase over the next few years. Your decision is also influenced by the TV you will use. If your unit is not capable of HDTV and you have no intention of upgrading in the near term, buying an HDTV capable receiver might not make economic sense. However, you should keep in mind that the HDTV feature is certainly not the only decision point you need to consider when making an informed choice regarding the optimum FTA receiver for your needs. FTA is not yet mainstream, and units are not available in every electronics store as are more common electronics items such as DVD players.

The sections that follow offer baseline considerations for purchasing an FTA receiver. They apply equally to a new or used unit. Instead of the worn-out cliché, "Buyer beware," I like the phrase, "Buyer be aware"—that is, know what you are buying and why. It is the best way to avoid any pangs of buyer's remorse after a purchase.

This chapter will help you become an informed buyer who is capable of choosing a receiver unit that will meet most, many, or all of your requirements for a first-time FTA receiver purchase. And with any luck, you won't have to break the bank when you buy it. The sections that follow offer information that will help you understand a little bit about many of the most important features found in FTA receivers. Read this section before you review the specifications or features list for any receiver you are considering. Much of the information will be covered again and in more detail in future chapters of this book, where the discussion will address putting the features to use. The explanations here are brief and uncomplicated.

Expect to learn just enough about some fundamental specifications and features as you consider which receiver to buy.

NOTE *For a list of potential sources for FTA satellite receivers and equipment, see Appendix A.*

Preferences

It is important to differentiate between preferences and technical specifications. Preference elements involving a purchase would rarely have bearing on the specifications or performance of an electronics purchase. Case color as an example would have no relevance in regard to how well a receiver worked. Intangible factors, somewhat harder to quantify or compare, can have an impact on how satisfied one may be with any electronics purchase. So first in the next two sections the text will consider two of the intangible preferences that may apply to FTA purchases.

Brand

As a purchaser of appliances, electronics, computers, or automobiles, you have probably developed on your own or learned a brand preference for the items you intend to buy. This brand preference also carries through into the consumable products we buy and use every month, such as soda pop or laundry detergent. Brand preference is strongly influenced by the volume and quality of advertising that is it on TV or radio, in printed matter, or through word-of-mouth ads for the product.

With FTA currently considered a niche market in the United States, few if any local dealers are currently peddling the products, media advertising for receivers is virtually nonexistent to few and far between. So I think with purchasing any FTA receivers for the first time, your brand preference will be muted at best, at least for most buyers in the United States.

As an informed buyer, you can take a methodical approach to the issue instead of being overly concerned about brand. Across the nearly 40 or so available brands you could purchase after checking for actual availability and pricing, first stay focused on the few important technical details. Next consider the warranty periods and feature-for-feature comparisons for the price between models and brands. This will likely lead you to the top selling brands, as long as other FTA consumers are doing their pre-purchase homework, too. If this does not result in finding the top selling brand in that category, it should at least help you find the best buy at any given price point.

Dealer

Does where you buy from matter? You could easily choose to buy a receiver or any product simply based on price alone. It is only natural to want to grant your business to a provider of products and services with the lowest prices. Where you buy the unit potentially will make a difference, perhaps a big difference if there are any problems with the unit. The dealer's return policies, restocking fees, free and fast in-kind replacement for DOA (dead on arrival) equipment, technical support availability, available repair service, or any value-added bonus items are not necessarily tied tightly to price anymore. The Internet has driven so many items to commodity type price-only purchases.

Seems that on the Internet stores, every item for sale is treated the same as a box of wooden pencils. Anybody could sell you a box of pencils and you would probably be satisfied with your purchase. With a lot of equipment sold and purchased over the Internet from multiple sellers, prices tend to find the rocky bottom, thereby limiting what can be included as value-adds with the purchase. Dealers that have managed to get the sales volume up while keeping their prices relatively low can have an advantage of offering some of the intangible elements along with the sale.

On any purchase, you will be taking a risk regardless of where you buy. My advice is to shop price first. Always compare same unit/model to same, or at least feature set to feature set, and at least know the post-sale risks based on the seller's advertised policies before you buy. Try to choose a dealer with some time in the business or at least one that is focused entirely on and serious about the sale of FTA equipment and not a dealer that also sells shoes or knockoff purses. When ordering from Internet sources, be sure the items you order are in stock and ready to ship.

Receiver Features

The purchase criteria in the next sections are feature specifications that can be compared with a simple check mark that each unit under consideration will either comply with or not. As you read through these sections consider making a list of the items most important to you.

Operational Worldwide

At first you might wonder why worldwide operations would matter in your selection of FTA satellite receivers if you are not a frequent worldwide traveler. Think of it this way: regardless of the country you are in now, you are likely to want to watch shows broadcast to your location via FTA from another country's satellite programming. This feature, when stated in product literature, implies that the unit meets the international standards that are at play in the FTA satellite technology environment. This benefit implies some, if not complete, compatibility not only with the majority of channels and programming but with the peripheral video and sound equipment or television sets sold worldwide. Essentially, FTA knows no borders and its only "barrier" to your getting and using the signals is the curvature of the

Earth itself—or perhaps a neighbor's mighty oak tree that blocks the view of the southern sky preventing a good signal.

On-screen Menu Display with Multilingual Support

Theoretically on-screen menu displays (OSDs) could use any known written language. Units for the North American market will typically be able to display the menu's wording and numbering in English, French, and Spanish at a minimum. Normally, multilingual units will also include choices for Danish, Dutch, Finnish, Greek, German, Italian, Swedish, Portuguese, and Turkish languages. More languages are possible; just be sure that yours appears on the unit or one you can understand.

Processor Speed

In many ways, the satellite receiver works and functions behind the scenes much like a computer that is designed and programmed to perform only the necessary functions of receiving FTA signals, decoding the video and audio data that is received, and outputting that to a screen and speakers or to a recording device. It is a computer with a limited scope of work to do. From the user's perspective, the easier the interface (remote control and menu system, not a keyboard) is to operate, the better. A higher processor speed helps to perform menu functions quickly and speeds up routine tasks such as channel changes or background tasks. Generally speaking, the faster the processor speed, the better the experience will be for the user of the device.

Operating System

Technically oriented readers might have a preference for one operating system (OS) over another. The OS makes the receiver's computer chips work together to get the video to the screen or recording device and performs all the background chores, such as collecting and storing programming data or operating on-screen menus. For most users, the OS will not come into play, and their only use of the OS will involve interfacing with its preprogrammed commands via the remote control. The key questions for the casual user will simply be these: Does the device perform as promised and does the potential exist for upgrading the unit's operating system when needed?

Various versions of UNIX and Linux are common in FTA receivers, and unless you are fluent in using and programming these OSs, it probably will not matter which one the receiver unit uses. The factory OS upgrades from the manufacturer are usually sufficient for staying current and eliminating any "bugs" discovered in the OS. The upgrades could occur automatically if the unit is connected to the Internet; otherwise, upgrades can often be loaded from a computer over a USB port.

NOTE *Attempting to make changes to the programming or OS not approved by the manufacturer is ill advised. Altering the OS or the operation program can also alter the warranty or degrade the performance of the unit. If it works, let it work.*

Standard Definition or High Definition

Standard definition is a given criteria—a must-have on any FTA receiver. If you need or want to receive HDTV, the receiver will have to support that mode. HD (720p resolution) channels are not the most widely available or even the "standard" for FTA satellite transmissions. Most programs are available in standard definition (SD). HD is so much better for the programs that do broadcast in HD, and more stations will probably broadcast in HD in the future, so having this feature might be worth the incremental cost.

Digital Video Broadcast Receiver Decoding

Digital Video Broadcast Receiver Decoding (DVB-S) is the most common physical layer format for video and sound data transmission over satellite. Any receiver you purchase will have to work with DVB-S. DVB-S2 is simply the new and improved version of DVB-S that provides the path for higher definitions and advanced audio processing.

MPEG-2 is a format for encoding video and audio in any digital transmitting and receiving environment. To receive and display MPEG-2 content, the receiver must support and "decode" this data stream to your screen and sound system or to and from your personal video recorder (PVR) feature.

MPEG-4 is an improved version of MPEG-2 but is divided into component parts, or stated another way, a collection of subordinate standards—28 in all. Look for compatibility with H.264 at a minimum.

Supported Screen Formats

A receiver claiming "Auto" as a supported format means it will switch aspect ratios to match the incoming signal. A 4:3 format display is similar to that of old analog television sets, and a 16:9 aspect ratio is a widescreen format. The best rendering of a video stream occurs when the display modes of the receiver and the TV match exactly the display mode in which that video stream was transmitted. A receiver with multiple display modes will render video outputs in 720p, 570p, 567i, and 480p, and the newer ones will include 1080i.

Audio Output

For analog audio, composite video/audio output connections on a receiver will handle mono and stereo level broadcasts and transfer the sound signal to your player or TV. Analog will be output through an RCA jack and will have left and right side sounds for stereo sound.

For digital audio, to handle 5.1 digital sounds, the receiver has to support six-channel audio, noted as AC3, and have a jack to output that to your sound system. Digital audio output or S/PDIF output will use a 75-ohm coaxial cable.

There is a trend to use fiber-optic output to carry the digital sound signals, but traditional wire cables can handle the signals as well. If you have an advanced sound system, check to be certain that the output/inputs jack types match. If you have no plan to use an advanced sound system, you can still listen to the programs and movies in mono and stereo modes using the analog composite connection to a TV.

Receives SCPC and MCPC

The satellite receives its signal from a ground station and retransmits the signal on a transponder within a 27 MHz frequency band. There are two ways the transponder frequency bandwidth can be used. The first way, multiple channels per carrier (MCPC), time slices the bandwidth and transmits multiple channels of data (video and audio). The technology that enables MCPC is called time division multiplex (TDM). The channel's data is broken into small pieces, and each respective received channel's data is received in sequence, one behind the other. The second alternative to use the bandwidth is single channel per carrier (SCPC). Both the C and Ku band transponders use both of these transmission methods based on the ground transmitting station's criteria.

LNB-Switching Control

To receive any signals at all from a satellite, the dish has to have at least one LNB to connect to a receiver. Having one LNB receiving from one satellite limits what programs can be received from that satellite. The solution to improve the number of satellites that can provide programming to a receiver is to use more than one LNB. The technical standard that defines the ability to manipulate which LNB is sending the signal from a satellite dish when more than one LNB or when more than one dish is connected to a receiver over the same cable is referred to as DiSEqC1.0 and 1.1.

A *4 in 1 switch* device allows a connection to more than one LNB operating on a multiple LNB satellite dish. This switch can also be used to deploy more than one dish, each with its own LNB, with the additional dishes aimed at other favorite satellites. Typically available switches are units for either switching between two LNBs or for choosing one of four LNBs for originating the signal. The LNBs are cabled to the switch's inputs, and one cable goes to the receiver from the switch. As channels are changed on the receiver requiring a different dish or LNB, a tone is sent to the switch that "rotates" to the called-for LNB input, sending its signals only down to the receiver. This convenience feature is necessary if you intend to use more than one LNB or more than one dish to look at programming from more than one satellite. How to hook up multiple LNBs and multiple satellites is covered in more detail in Chapter 8.

Motor Support: DiSEqC 1.2/1.3 (USALS/Go to X)

There are only a few ways to locate a signal from a particular satellite. The first, of course, is to set in the azimuth and declination angle manually so the dish antenna is aimed properly to receive the signals from that satellite. When you want to move to a different satellite, it is possible but very inconvenient to go out to the dish and re-aim at another satellite. The easier solution for moving from one satellite to another is to install a dish motor that will take instructions from the receiver over the cable to move the dish to point at a different satellite.

The motor's instructions are sent over the cable by a standard communication protocol called DiSEqC 1.2. Some receivers also support a little program referred to as Go to X or Universal Satellite Automatic Location System (USALS) also known as DiSEqC 1.3. With USALS or Go to X, quite simply, once the dish is set up to know south as a direction (or north in the Southern Hemisphere) and the installation latitude and one satellite is found, it can find the rest of the satellites by sending the control information to the dish motor. Looking at two satellites can work with a switch and two dishes fairly easily. Finding more than two is far easier and convenient with a motorized dish and a receiver that supports DiSEqC 1.2 signals and has the USALS (Go to X) program built in.

Video Outputs

High definition multimedia interface (HDMI) output or interface means that the unit has a connector to use the HDMI device interface, which is a 19-pin connector that carries both digital (multichannel) audio and uncompressed video signals from one HD-capable device to another (FTA receiver to television). This is a preferred type of connection for modern HD devices. The latest standard improvement to HDMI is called HDMI 1.3, which will vastly increase the future possibilities for higher quality video and audio.

Other possible video outputs are the following:

- **YCbCr** A digital video signal standard that uses one cable
- **YPbPr** An analog video signal that uses three cables, with RCA jac plugs

- **S-Video** A two-channel video signal that typically uses a 4-wire cable with a mini-DIN connector
- **Composite** The older single cable method of transferring a video signal from one device to another over a single cable

The important thing to consider is matching the FTA receiver output to an available input on the TV or alternatively to a recording device. The next consideration is whether the receiver will provide video signals to the TV over the best possible connection standard for that particular TV.

PVR

The FTA receiver decodes the video and audio signal for the channel you want to watch and would normally route that decoded format (the program you want to watch) to the screen for instant viewing. With a PVR feature, also called a DVR (digital video recorder), the program can be stored for later viewing instead. The saved program file can be stored on an internal disk built into the device or an external storage device can be used. The USB (universal serial bus) standard provides a method to plug in jump drives or portable hard drives if the receiver does not have its own internal storage. This feature can be used to play compatible video/audio files on the jump drives that originated from other sources. The price of the receivers is kept low by not including hard drives in the units. USB drives can be purchased separately in many sizes and prices.

Seven Days Electronic Program Guide (EPG)

This feature allows the receiver to track and display up to a week's worth of program information and display it as On-Screen-Display (OSD) when the telecaster provides this information to be displayed. Newsfeed channels receivable on FTA from major networks' news-gathering efforts, for example, will not provide this data, as they have no way of knowing ahead of time what will be on the newsfeed channels.

Channel Scanning

Once the dish is pointed at the desired FTA satellite, the next trick is to find all of the available FTA channels on all of the transponders (retransmitters) emitting signals from that satellite. These channels can be added to your channel/program list in two ways: You can enter the data manually, or the device can do all of the work by scanning all the possible frequencies and making the entries automatically. This is referred to as blind scan. Manual scan can be useful. For example, if you have already done a scan and you find out a single new station is on the satellite, it can be added via a manual entry scan. Automatic channel scanning support for both SCPC and MCPC transmissions types and DVB-S and DVB-S2 encoded signals would be highly desirable.

Videotext Decoder

"Videotext" and "teletext" are the terms used to describe text-based information that is included with a video broadcast and projected in text to the screen, usually at the bottom. Closed-captioning is a derivative form of videotext. Important news of the day can be broadcast via videotext.

Upgrade Software

Periodically, the manufacturer of the unit can make available software upgrades to the operating system, firmware, or user interface software package features. The two most common upgrade interfaces are via a USB port, where you download the upgrades to a computer and transfer them, or you can directly receive updates via the Internet through Ethernet port connected to a router via a WI-FI connection to the Internet.

Parental Lock

The parent lock feature allows you to set a password before a user is allowed to view certain content.

Card Readers

A smartcard reader is less important in the North American market for FTA broadcasts. The major pay-for-TV satellite providers want you to use their equipment, not FTA receivers. Smartcards let you view paid-for programming, which is encoded (scrambled), and the smartcard provides the information to decode (unscramble) the transmission. In TV markets in some countries, the pay providers and the FTA providers play better together, and in some markets an FTA receiver can receive the paid-for and premium channels decoded with a smartcard that is inserted into the FTA receiver. Two versions of smartcard readers exist: common interface (CI) and certification authority (CA).

Internet Interface

Receivers that can get upgrades over the Internet require an RJ-45 jack to connect to a router or Internet modem.

SCART

SCART is an older device interface used on European video and audio equipment and on TVs.

Features: The Bottom Line

Do not let the laundry list of potential features overwhelm you. Some of the advanced features will be transparent to you when you do your receiver setup, and you might never use others. Stay focused on the most important details. The most important features are fairly straightforward: The video and audio has to match your equipment for viewing and/or recording. Essentially, your TV and advanced sound system, if you have one, must interface via the connection you have available.

As a novice user of the receiver, automatic channel scan is an important feature. Manual scan (setup of channels) will use up a lot of time. Having the ability to look at more than one satellite will be of value in increasing your viewing opportunities whether you choose to do that with a multiple LNB dish, a motorized dish, or multiple dishes, so either (or both) switching and motorized satellite locating should be supported on the receiver to do this. If you want to catch shows while you are away to view later, PVR support on the receiver is a necessary feature.

CHAPTER 7

Getting the Signals: Aim and Synchronize the Dish

Think about it: you're pointing a parabola-shaped dish toward the sky, hoping to align the center of the dish with a small satellite floating in space about 23,000 miles above the equator. It seems like a nearly impossible undertaking. Fortunately, communications satellites that relay television signals back to Earth are maintained in an orbit position relative to the Earth that does not change much. This path is called a *geostationary orbit*. To receive the signals from any one satellite, you need to aim the dish as close as possible in the direction of the stationary satellite.

If you are standing anywhere in the Northern Hemisphere, facing the North Pole, the sun appears to rise up from the horizon to your right, or from the east. We say that the sun rises in the eastern sky and sets in the western sky; however, what is really happening is that the Earth is constantly rotating on its north-south axis toward the east at a rate of one complete revolution per day.

For a satellite to maintain a nearly constant position above the Earth, it too must orbit toward the east—that is, it must move in the same direction that the Earth rotates on its axis. The satellite's speed of movement must be sufficient to complete a rotation around the Earth's north-south axis at exactly the same angular position to a point on the Earth's surface. To maintain that position, it must travel at speeds close to 7000 miles per hour. The satellite's track is maintained on a parallel path to the Earth's equator.

Fortunately, it is not necessary for us mere laypeople to do all the mathematical calculations necessary to determine aiming angles. This satellite aiming information is available from a number of sources, including interactive web sites. The geometric challenges surrounding dish aiming are the same for paid satellite TV as with FTA, except that LNB skew will not be an issue with circular polarized LNBs. DIRECTV, DISH Network, and Bell are known to use circular polarized LNBs. With paid-for satellite providers, the setup routines include alignment data that work by your entering ZIP code locations that are shown on the receiver's setup screen, or you can use the online lookup tables; or you can call the provider's customer service center and give them your location, and they can provide the azimuth and elevation data you need.

With FTA you have to look up the information on your own to align the dish antenna properly for a particular satellite. When setting up a motorized dish for the first time, you need to align the dish to the nearest southern satellite. It is helpful to have a good understanding of the underlying geometry involved in getting the dish into proper alignment with the incoming satellite signal. The next sections provide the contextual basics that effect the alignment requirements.

Getting the dish into alignment might be the hardest part for many beginners. Have some patience, check your plan and numbers twice, and then proceed with the setup step by step, checking your work at each point as you go through the steps. If working above grade, keep that safety cap on and use all necessary safety precautions.

Earth and Satellite Geometry Basics

The challenge to receive signals from a particular Ku band satellite is to align an imaginary straight line from the center of the dish directly to the mass of the satellite. Fortunately, there is an allowance for not being perfect and exact, which means that you just have to be very, very close. To align the dish to the desired satellite's signal, you need to know the fixed location of the satellite in its orbit and the longitude and latitude of the receiving dish location on Earth.

Pinpointing Receiving Locations

The best way to describe a fixed point on the Earth's surface that can be universally understood and located is to use the angle measurements of the degrees of a circle. Each degree of the compass angle is divided by 60, with each of these divisions called *minutes*. Each minute of angle is further divided by 60 and is referred to as *seconds*. So degree measurements are used in two planes—*longitude* and *latitude*. Longitude lines divide the Earth vertically from pole to pole, and latitude lines divide the Earth from the equator to both poles.

Degrees of longitude are imaginary lines that subtend out from the north rotational pole to the south rotational pole with the 0/360-degree line, called the *prime meridian*, passing through the town of Greenwich, England. Each successive meridian or longitude line from 0 to 180 is drawn from the prime meridian, with one set going west and the other going east. It's as if the degree scale of a protractor was placed at the poles and a full half of the circle degree marks were made at the equator going in both directions, starting from the prime meridian. These successive degree marks are referred to as their *angle of degrees* west or east, from 1 to 180. Notation for halfway west would be 90.0° W.

The 180-degree position in the western Pacific Ocean is called the *International Date Line*. Going east of the prime meridian, the degree of longitude at the halfway point is annotated as 90.0° E. The 180 degrees east and west line of longitude is the same line, the International Date Line. At the equator, with the sun directly above the equator at summer solstice, the Earth rotates at the rate of 1 degree of longitude every 4 minutes relative to the sun. When using decimal mathematical notation for west longitude, the number is preceded by a minus sign. Going east of the prime meridian, the numbers are positive.

Latitude lines are set from the equator, with the equator itself being the 0-degree mark. In measuring latitude, it's as if the center of the Earth became the setting point for the dial of a protractor, with 0 set at the equator and 90 degrees at each pole. Each degree mark creates an imaginary line circling the Earth that is parallel to the equator. The lines north are annotated as 1° N, 10° N, and so on. Degree lines of latitude south of the equator are followed by an S, such as 45° S. This line could also be called the 45th parallel, but that alone will not pinpoint a location, because there are two of these lines. Hence the need for S and N annotations to differentiate between latitude lines south of the equator or north of it. When expressed in terms of mathematical notation, latitudes north of the equator are positive and latitudes south are negative values.

Locations on Earth can be precisely located and universally understood using this global coordinate system. The annotation for a location can be expressed in degrees, minutes, and seconds or entirely as degrees with decimals for values less than a whole degree. The latitude (N or S) is customarily listed first, and then the longitude (E or W). So, for example, the annotation for the area of the parking lot behind Ron Jon's Surf Shop in Cocoa Beach, Florida, can be expressed as follows:

- In decimal degrees using mathematical notation: 28.3565, –80.6072
- Or in degrees, minutes, and seconds: N28° 21' 23" and W80° 36' 25"
- Or in degrees and decimal minutes: N28 21.390 W80 36.432

NOTE *All three annotations refer to the same general location. All three are sufficient for accurate dish aiming.*

Remove the minus (–) sign on the 80.6072, and you have travelled east along the parallel to a point just outside of New Delhi, India, close to the border with Nepal. Remember that a minus sign on the latitude means the location is south of the equator, and a minus sign preceding the longitude means that the location is west of the prime meridian.

When converting readings from a map that are in degrees, minutes, and seconds to decimal, the formula is as follows:

$$\text{degrees} \times 1 + (\text{minutes}/60) + (\text{seconds}\ /360) = \text{decimal degrees}$$

So using the latitude from earlier to convert to decimal would look like this:

$$28 \text{ degrees} \times 1 + (21/60) + (23/360) \text{ yields } 28 + 0.35 + 0.006388 = 28.356388.$$

Going the other way, the formula is reversed. Whole degrees = whole degrees. The remaining decimal is multiplied by 60 to yield whole minutes. The final remaining decimal is multiplied by 60 to yield whole and decimal seconds. Seconds must be rounded up or down to the nearest whole number.

Here's an example:

80.6072. 80 degrees = 80 degrees

$0.6072 \times 60 = 36.432 = 36$ minutes

$0.432 \times 60 = 25.92 = 26$ seconds

80° 36' 26"

More traditionally, we have identified the location of our place of residence as a particular house number and street, city, and state. ZIP codes were later added to that address scheme. To use the various online lookup tables to find particular satellites aiming data, begin by entering the satellite and then the receiving location by address, by ZIP code, or by coordinates. Even when you have more traditional address info available, using GPS coordinates results in getting the most accurate dish-aiming data. A GPS and even some cell phone applications can also be used to achieve GPS coordinates within a reasonably accurate range.

Pinpointing Satellite Locations

Satellite locations are pinpointed using references to longitude as well. Their location above the Earth's surface is fixed to a circular orbit with a radius of approximately 30,162 miles from the center of the Earth, tracking along the line of the equator. Their rotation speed is just enough to keep them "tied" to a projected longitude line. Their orbit keeps them at the equator above a given longitude line at all times. One of the popular satellites for FTA viewing from the United States is fixed at 97° W; another is located at 125° W.

From any location on Earth, only a portion of the satellites in orbit will be within possible receiving range. Figure 7-1 shows a representation of the Earth with the surrounding equatorial geostationary satellite belt surrounding the planet. The dots representing the satellites in the drawing are 6 degrees apart. Because of the curvature of the Earth, no matter where the receiving station is located, your dish will not be able to see out-of-range satellites, because the Earth itself gets in the way. The in-range satellites, using the Earth's center as the radius point, will typically be within 144 degrees of longitude, as shown in Figure 7-1. The solid west limit line and the east limit line mark those points. If you move the receive point anywhere on the planet east or west, similar limits apply. The long-dash short-dash lines are impossible—one is quite obviously out of sight. This limit of range is also reflected in the dish motor's ability to rotate the dish. Typically dish alignment motors will be limited to rotate + or – 65 to 70 degrees of rotation from the true south or true north setting.

When you're using any of the web sites for interactive satellite location tables, if the results are in negative numbers for *dish elevation*, then the Earth is in the way and it is not possible to receive signals from the subject satellite. The same problem occurs from high latitudes. If the receiving location is near or above 80 degrees north or south latitude, then the Earth's curvature is again a problem for reception. Remember that any time the dish elevation setting is in negative numbers, the satellite is out of sight range.

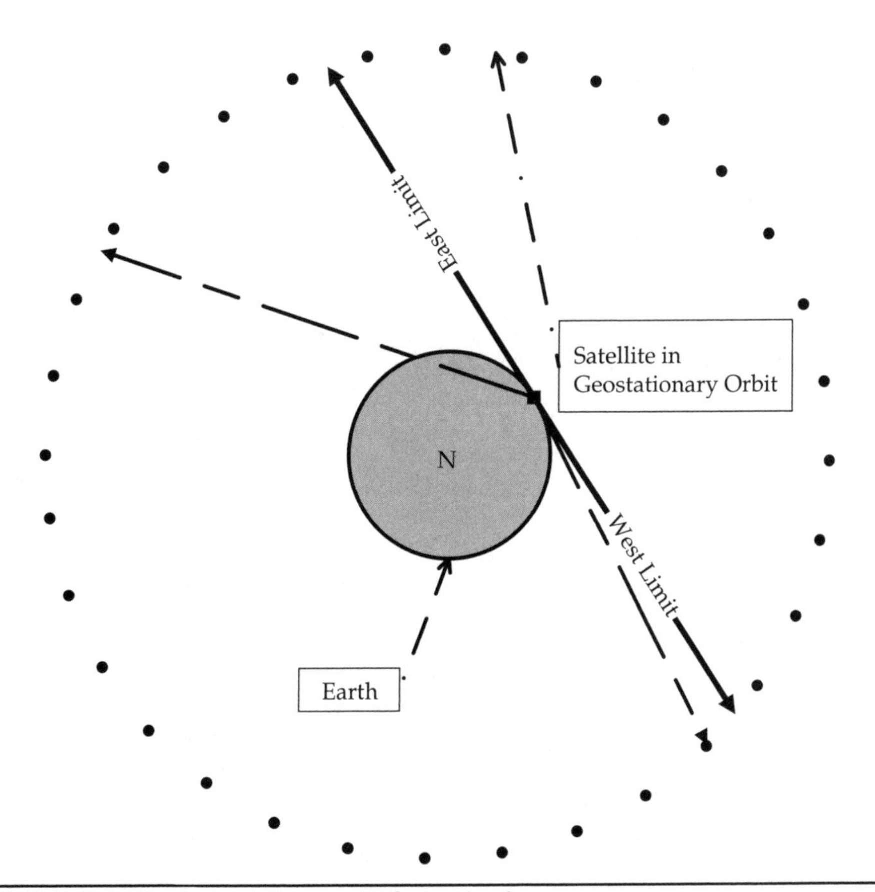

FIGURE 7-1 Satellites in geostationary orbit from polar view

Figure 7-2 shows a representation of geostationary satellites in orbit from an equatorial viewpoint. The dashed line passes through the tangency point of the Earth's horizon, making the satellite out of view and out of receiving range of the signal.

Dish Aiming

It is possible to aim the dish for proper azimuth, set the elevation, and set the required skew with simple, inexpensive tools. You can spend money on expensive aids, which are nice to have but are not entirely necessary. Some aids are in kit form with signal strength meters with portable power supplies. Compasses are available that clamp on the dish to assist with alignment.

The rest of this chapter will walk you through this process, requiring only some knowledge and a few hand tools. In this example, the notion is that the dish is either ground-mounted or portable. You will have to improvise the azimuth procedure only slightly for a building-mounted dish. The essential items are listed next.

Tools and Information

You'll need the following tools to aim the dish:

- Magnetic compass
- Three wooden stakes

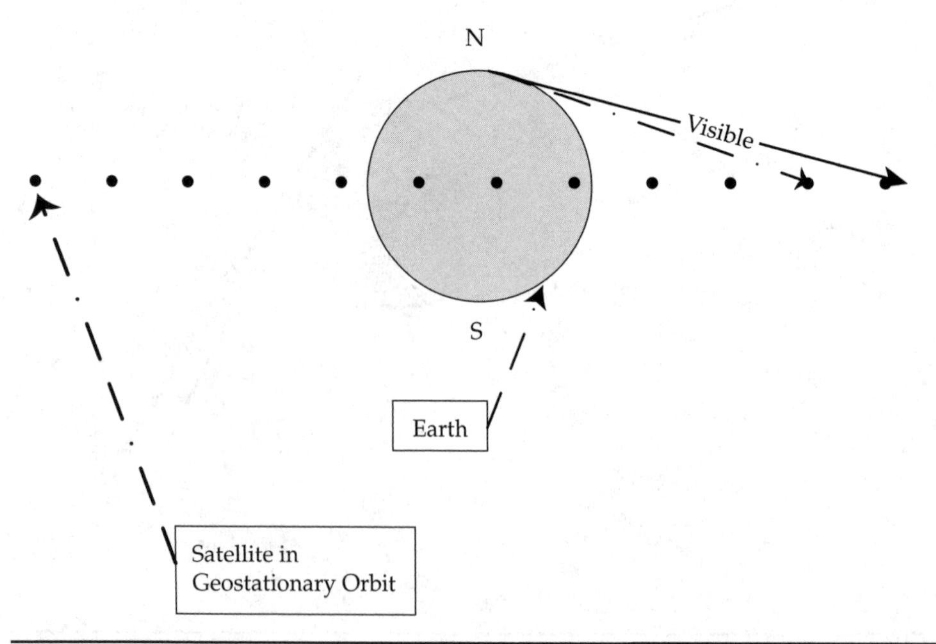

FIGURE 7-2 Equatorial representation of satellites in geostationary orbit

- String
- Plastic protractor
- Small weight (fishing, nut, or bolt)
- Heavy sewing thread
- Phillips screwdriver
- Torque or box wrenches (10 mm)
- Small bubble level

You'll need to know the following information unique to your location to set up the dish to aim at a given satellite:

- Magnetic azimuth
- Dish elevation
- LNB skew (for linear polarized signals)

Choose a Lookup Table

Choose one of the many satellite aiming interactive web sites and enter your location and select the name of satellite you want to view. The satellite can be referred to by its title or its longitudinal position. (See Appendix B for URLs for lookup tables, or visit the author's web site at www.allaboutfta.com for the latest links to lookup tables.) The lookup table will return either three or four types of information: elevation, LNB skew, the magnetic azimuth, and possibly the true azimuth.

Using the Galaxy 19 Satellite at 97° W and given a receiving location of 28.2367, –80.6147, the lookup table returns the following information for our example:

- Elevation: 52.4 degrees
- Azimuth magnetic: 216.5 degrees
- Azimuth true: 211.9 degrees
- LNB skew: +27.7 degrees

The nearest "south" satellite from this given location for a motorized dish setup would be the satellite Nimeg 4 located at 82° W. The returned table information for Nimeg 4 is shown here:

- Elevation: 57.0 degrees
- Azimuth magnetic: 187.5 degrees
- Azimuth true: 182.9
- LNB skew: +2.6 degrees

The difference between the true and magnetic azimuth is the magnetic declination angle that recognizes the difference between the Earth's rotational pole and its magnetic pole. The magnetic north reading on a compass changes from location to location.

When using lookup tables for satellite locations, where postal address information is hard to come by or potentially inaccurate, such as remote cabins or when traveling in an RV, it's best to use coordinates from a GPS if you have one. Some interactive map web sites will also allow you to pinpoint your location and obtain the coordinates.

In the examples that follow, we will use the settings for the 97° W satellite from the dish receiving location of 28.2367, –80.6147 south of Cocoa Beach, Florida.

Set the Dish Elevation

Start the process of linking to the satellite signal by setting the dish elevation. Dish elevation identifies how high up into the sky the dish must point to meet the incoming signal of the satellite head on.

1. Verify that the mounting pipe is vertical (plumb) in both dimensions: check front to back and side to side. Both must be vertical, perpendicular to the horizon. The bottom bubble on the level shows that the pipe is vertical when the bubble is between the lines on the bubble. If the mounting pipe is tilted off level, the readings for elevation on the butterfly flange cannot be accurate. First check front to back in line to the target satellite as shown here:

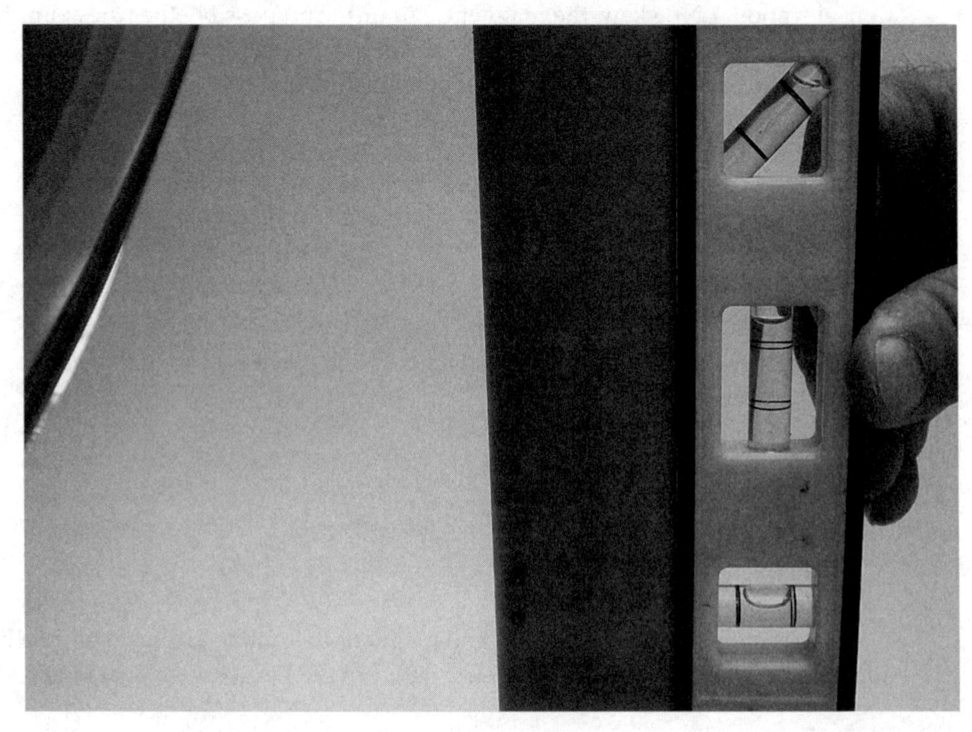

Then check the pipe side to side parallel to the dish face as shown, again looking at the bottom bubble on the level:

If the pipe is not perfectly plumb, adjust the mounting bracket, the tripod if you are using one, or the mounting situation to achieve plumb level readings on the mounting pipe both front to back and side to side. You have to adjust and correct whatever is necessary in the situation to have the mounting pipe perfectly vertical.

2. Once the vertical mounting pipe position has been verified, you can set in the dish elevation. As shown in Figure 7-3, the required elevation for the 97° W satellite based on the receiving location is 52.4 degrees. The setting for 0.4 requires a calibrated eye, because tenths of degrees are not marked on the elevation, and every tic mark on the butterfly flange's elevation scale represents 2 degrees. The human eye is capable of discerning fine gradients to a limit of about 0.01 inch. In this case, to achieve the 0.4, set the indicator halfway between 52 and 54 degrees and back it off from halfway just a little bit, as shown, achieving an approximation to 0.4 degrees.
3. With a box wrench or socket wrench, tighten and torque all the setting nuts to very snug. If you have a torque wrench, set it for 20 foot pounds or read that on the dial. If you do not have a torque wrench, tighten to a point where the nuts are tight, then make a quarter turn more without applying excessive force. Applying excessive force can snap the bolts in half or bend or fracture the flanges or pipe.

After the elevation is set, your next step is to set the dish along the proper azimuth.

FIGURE 7-3 Setting dish elevation

Set the Dish Azimuth

From any receiving location, the satellite's location is identified by the angle (or direction) that the dish has to be pointing to achieve a straight line from that point on Earth to the satellite. From the lookup tables, the magnetic azimuth identifies the angle the dish must point in reference to compass heading relative to the Earth's north magnetic pole, considered to be the 0 degree point.

The magnetic azimuth is a line projected along a trajectory out from a degree reading on a compass dial. Any one of three types of magnetic compasses can be used to align the dish. Two of the better choices are a lensatic compass, sometimes called an engineer's compass, or an orienteering compass, sometimes called a map compass or protractor compass. The compass north-seeking needle points to magnetic north, and the azimuth in degrees is read off the compass dial. When using the engineer's compass, lay-in an azimuth line looking through the slot and align that with the wire, then match the wire to a distant location. The orienteering compass is clear plastic and has a pointer on it to line up with the desired azimuth.

Figure 7-4 shows the simplest oil-filled magnetic compass with the north needle of the compass pointing north and the case dial set to 0 degrees as north. There

FIGURE 7-4 Ordinary compass used to set in azimuth line

is one easy way to identify the magnetic azimuth. In Figure 7-4, a mechanical pencil lead is set to the desired degree reading of 216.5 degrees using the underlying degree dial while the compass is set to north and the dial is set to 0 under the north-pointing needle. Using this method, you would look down the pencil lead line to find a point in the distance that is exactly on the imaginary line extending out at the 216.5-degree azimuth. The tip for doing this is twofold: avoid having the compass close to metal objects and do not hold the compass in your hand. Set the compass on a stable, nonmagnetic resting point.

Regardless of which compass you use, the process is the same. The dish has to be pointing out along the desired magnetic azimuth. The mistake that is often made by novices is to use reference points that are too close, such as the bracket arm on the dish holding the LNB. The azimuth line can be laid-in in a most accurate fashion if the reference points are at a distance.

1. Select a point behind the dish location (if possible) and set a stake as the starting point.
2. Use the compass alignment on the correct azimuth from the first stake to set a second stake where the dish will be located. The dish mounting pipe can serve as the second stake, when the dish is on a permanent mounting pipe.
3. Set a third stake along that same azimuth line, lining up the compass reading, to the first and second stake.
4. Go to the third stake and shoot a back azimuth to be sure the three stakes or reference points are perfectly lined up. A back azimuth is calculated by adding 180 degrees to azimuths less than 180 and by subtracting 180 if the original is 180 or greater. This seems like a bit of effort, but the whole process can easily be done in less than 15 minutes.

5. To assist with eyeballing the dish to point along that established magnetic azimuth, run a string from the second to the third stake. Any variation from this can work, such as lining up with a tree, fence post, or rock as the third point, as long as it is a narrow reference point and is exactly along the desired azimuth for pointing the dish.
6. Point the dish along that azimuth line as exactly as you are able. From the center top of the dish, an imaginary center line crossing the LNB must point exactly down the line to the third stake reference point.
7. Remove the stakes when you are done.

NOTE *A mistake that is often made is trying to use the south arrow for shooting in the azimuth line by rotating the compass dial by the amount of the number of degrees added to 180 to equal the desired azimuth. This does not work, because the south needle always points to a magnetic azimuth of 180 degrees.*

After you get the dish alignment to the desired azimuth, one more step is needed to capture the signal's full strength: tilting the LNB to match the alignment of the satellite's incoming signal is called *skew*.

Synchronizing LNBF Skew Angle

The need for setting linear polarization stems from the fact that the wave signal is composed of both magnetic and electrical elements. In linear polarization, where there is no phase shift between the magnetic and electrical elements of the signal, the signals can be either horizontally polarized or vertically polarized in reference to the Earth's horizon. With circular polarization used predominantly by C band and pay-for providers, there is a phase shift of 90 degrees between the electrical peak and magnetic peak of the transmitted signal. With FTA Ku band frequencies, it is necessary to skew the LNB so that its angle is synchronized with the incoming signal of the desired satellite's signal. If the receiving location is on the same longitude as the desired satellite, the LNB orientation angle is near 0, essentially even with the horizon. In the Northern Hemisphere if the satellite is westerly from the receiving location, the LNB rotation angle is positive. When the satellite is on a longitude east of the receiving location, the angle is negative. The X axis for the angle is vertical and perpendicular to the Earth and the rotation reference for counterclockwise or clockwise is standing in front of the dish with the target satellite at your back.

In the dish assembly in Chapter 4, the finishing step for the LNB was snugging it up vertically and centered front to back in its mounting bracket. Once azimuth and the dish elevation have been set, the next step is to align the LNB so that its internal antenna is spot on in alignment to the horizontal polarization of the signals from the target satellite.

1. Loosen the set screw on the LNB mounting clamp just until the LNB will move back and forth in its mounting and rotate freely within the clamp. The recommended LNBF (more commonly called LNB) rotation is expressed in negative or positive numbers. If you were standing in front of the dish with the target satellite to your back for negative LNB skew settings, the LNB is rotated within its mounting bracket clockwise. For positive settings, rotate the LNB counter-clockwise within its bracket.

   This rotation matches the horizontal center line of the LNB with the horizontal angle of the incoming signal. From any point on the planet's Northern Hemisphere, looking east the horizon falls away to your left if you're standing behind the dish, so the left side of the LNB's receiving head is lowered to match the slope of the horizon. If you're looking at a satellite more to the west of the receiving longitude, the right side of the LNB receiving head is lowered. When the signal is projected from the satellite's transponder at its longitude, the signal is even with the horizon at that point. As the receiving location is distanced from that of the satellite's own longitude, the Earth falls away in both directions because of the curvature of the Earth. Skewing the LNB brings the receiving element in the LNB back into alignment with the polarized signal.
2. The LNB and bracket likely will not be marked off in degrees to help with this setting for linear polarization. You can "eyeball" this for some settings fairly easily, such as 45 degrees, 60 degrees, or 30 degrees. Setting the skew for 11.7 or 35.5 might not appear to be that easy.

   You can make a simple and inexpensive tool to help, as shown in Figure 7-5. You will need two pieces of masking tape, a heavy washer or bolt or lead fishing weight, a plastic protractor, and a heavy piece of sewing thread to achieve more accurate skew settings. Figure 7-6 shows the skew angle being measured this way for a setting of +27.7 degrees; each small gradient mark on the protractor is 1 degree. You can still eyeball the 0.7 degrees as slightly less than 1, but more than halfway to the next mark.
3. Use the masking tape to secure the upside-down plastic protractor at the back center of the LNB. The thread with its weight dangling free should cross the 90 degree mark on the protractor and should line up exactly on the 90 degree mark on the protractor. The protractor's line from 0 to 180 should dissect the circular feed horn in half horizontally. An imaginary line from the centering point on the protractor and the 90 degree reading should dissect the center of the feed horn vertically.
4. Once the protractor is set to the LNB's horizontal plane, mark the top of the protractor's location with a light pencil mark on the back of the LNB. Place a vertical tic mark at the center point of the protractor to make realignment later easier if needed.

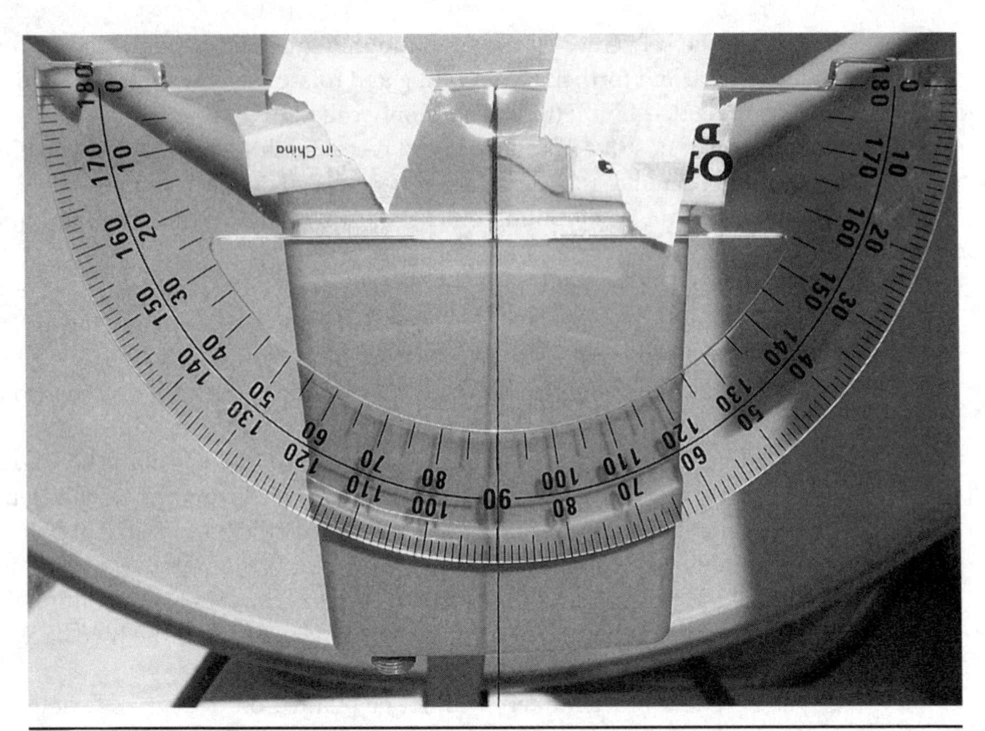

FIGURE 7-5 Homemade skew angle measuring device

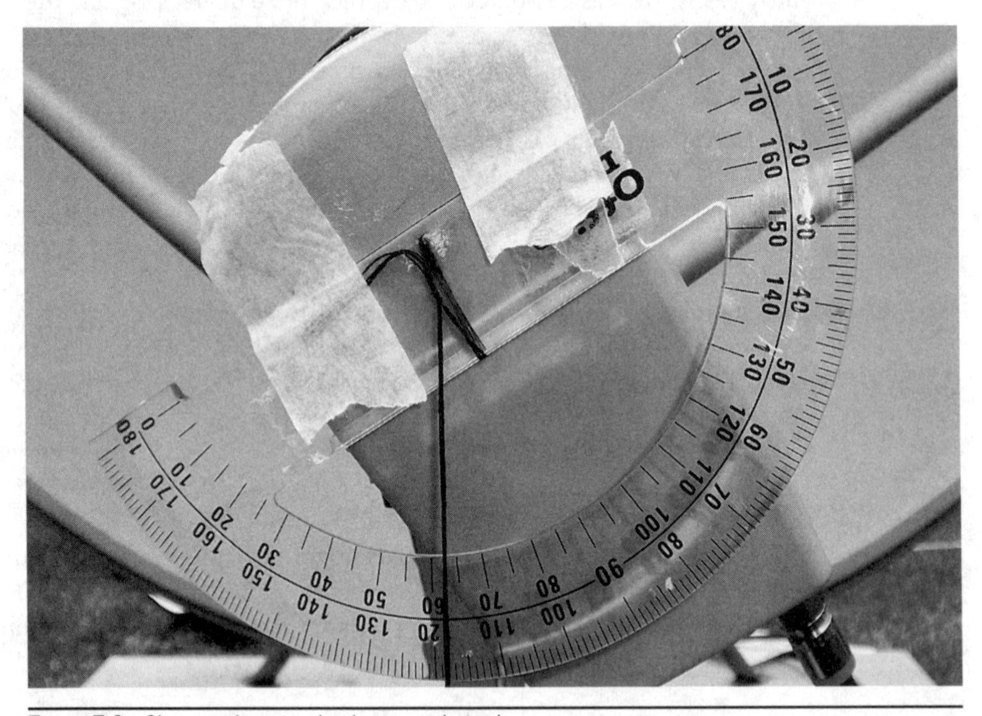

FIGURE 7-6 Skew angle set using homemade tool

5. Loosen the LNB set screw enough to rotate the LNB, and slowly rotate the LNB counterclockwise for positive rotation values and clockwise for negative values. Count the degrees of rotation as the thread passes the readings on the protractor. (Each tic on the dial is 1 degree. We are simply counting degrees off of the 90 degree mark. It's OK to do the math and add to 90 for counterclockwise rotation and subtract from 90 for clockwise, and then just rotate the LNB until the desired degree setting is reached. Either way is fine.)
6. Stop rotating when the desired number of degrees is reached, as shown in Figure 7-6.
7. Pull the LNB toward (front to back in the clamp) you and away from the dish while keeping its skew setting until it rests gently against the clamp. This sets the LNB at the proper distance from the center of the dish.
8. Tighten the LNB clamp set screw, being very careful to not move the LNB. If you find that you will be moving the dish often, say between two popular satellites with different LNB skews, you can pencil in the settings right on the clamp by marking a reference tic on the LNB and the two points on the clamp.
9. When you're finished, remove the protractor and tape. The only modification to the drawing protractor was to cut a barely noticeable notch with a pen knife for the thread to stay set at the reference point.

In all probability, the setting up of the dish to point at a satellite by going through all of the steps in this chapter can likely be done in less time than it takes to read this chapter. Once you have gone through the process, doing it again to select a different satellite is easy. When setting up more than one dish to catch the signal from another satellite, the process is the same. The only added advice is that during the setup of the second dish, do the blind scan of the new satellite by hooking up the down-lead to the receiver only on the second (or subsequent) dish, and then once the reception is verified, wire the new and original dish to the 4-in-1 switch. Four separately aligned dishes can be hooked up to a 4-in-1 switch.

Sometimes it might be necessary to fine-tune the dish settings while the system is on by putting a signal from one strong channel on the TV and watching the effect of moving the dish on the quality of the received video and audio signal. If you ever find this necessary, move the azimuth alignment by fine gradients first, then the elevation, and lastly adjust the LNB skew angle. Move the settings very slowly and wait a moment to discern the impact of each move.

The very nice thing about using a motorized dish is that the motor will tilt the dish to compensate for skew angle once it is set to the closest south satellite.

Motorized Setup

Setting up a motorized dish begins with following the motor manufacturer's specific recommendations for installation of the motor between the mounting pipe and the dish. The goal for a motorized setup is to align the dish to the nearest south satellite following the manufacturer's instructions.

There is a slight difference in the wiring diagram for using a motor to align the dish to available satellites, as shown in Figure 7-7. Power to drive the motor is provided by the receiver, so the RG-6 cable from the receiver is connected to the motor to a connection typically labeled REC. Nearby there will be another connector labeled LNB. You need to make a cable about 5 feet long with F-connectors on each end to reach from the LNB connection on the motor to the LNB itself. This slack is necessary because the dish will be rotated on its vertical and horizontal axes and the wire needs to be just long enough to accommodate this freedom of movement. The larger the dish, the longer the lead will be from the LNB to the motor. It is critical to use F-connectors that have watertight O-rings to seal out water on both

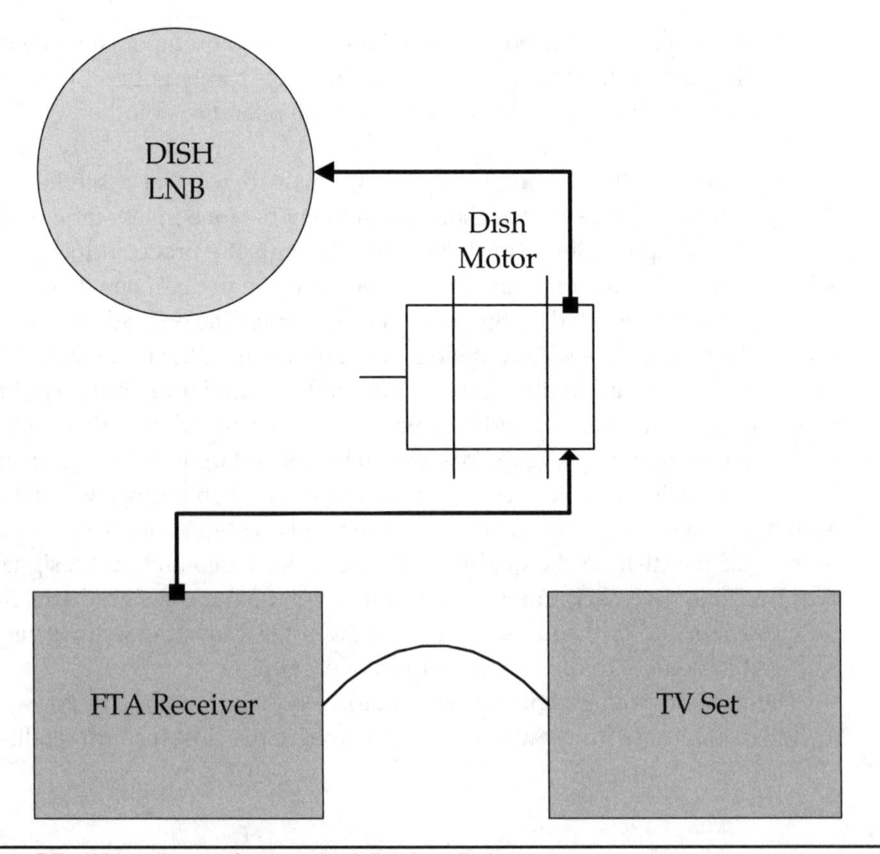

FIGURE 7-7 Wiring diagram for motorized dish installation

of these connections as well as the one connecting to the LNB. Alternatively, or even additively, a connector cover or "boot" can be used to provide moisture protection. This and any other outdoor connection will be the place where moisture and corrosion over time will potentially cause issues with signal quality or shorting of the motor drive voltage.

Two standards come into play for using a motorized dish. The receiver must support either DiSEqC 1.2 or USALS modes.

USALS is preferred because the setup requires alignment only to a true south azimuth followed by testing it with the nearest south satellite if you are in the Northern Hemisphere or the nearest north satellite if the receiving location is in the Southern Hemisphere. (USALS from a purely technical perspective is a program that calculates motor drive instructions and uses DiSEqC 1.2 instructions to drive the motor.) To achieve correct setup of a motorized dish for USALS, you must know your exact latitude, the magnetic azimuth from the receiving location to true south, and the name and location of the nearest south (or north for Southern Hemisphere) satellite.

After the dish is set for the latitude and set to the true south azimuth, the rest of the setup is done on screen with the receiver.

Select the proper mode (USALS) on the receiver using the remote control and then use the remote to enter the dish longitude and latitude and the name and settings for the true south (or north) satellite. The motor will then move the dish to find the near satellite set in. A one-time fine minor adjustment of the direction of the dish relative to the main mounting pipe might be necessary to optimize the signal quality. The features and screen fill-ins will vary from receiver to receiver. There can be preset satellite locations and learned locations, meaning others can be set in the receiver's memory as you use signals from different satellites. Once the dish is installed and properly set up, you can shift viewing from one satellite to the next simply by selecting the channel from your stored favorites list on receivers supporting this feature.

Dish motors will cost in the neighborhood of $100. Having one is not necessary to begin using and enjoying programming from FTA satellites. It is, however, a very nice upgrade that conveniently allows you to check for new FTA channels periodically on all satellites in range from your viewing location. Viewing programs from different satellites becomes very simple after the initial setup is completed.

CHAPTER 8

Using Multiple LNBs and Dishes

All the preferred FTA programming you might want to add to your favorite watch list might not be on the same satellite, so you will need to use one of three manageable methods to receive the signal from more than one satellite. One of the alternatives to moving your dish every time you want to tune in a different satellite is to deploy more than one dish or to use a dish with more than one low noise block (LNB) feed horn affixed to it to supply a signal to a single FTA receiver. To do this, the receiver must support the DiSEqC 1.0 or 1.1 (Digital Satellite Equipment Control) communication protocol. This protocol signal sent from the receiver toggles or rotates the compliant switch to access a signal from specific connection ports on the switch and sends only that port's signal along the down-lead to the receiver's input jack. The DiSEqC 2.0 protocol is a two-way communication that allows the switch to report status information back to the receiver. Many of the newer receivers also support the 2.0 protocol. With either protocol, the DiSEqC-compliant switch's job is to pass signals from multiple LNBs/dish antennas down to a single receiver.

As you develop a taste for more and more FTA programming from multiple satellites, make the choice early on as to whether you want to use multiple LNBs and multiple dishes or a single dish with a motor attached to the mounting pipe, as discussed in Chapter 4. To use a motorized dish, the receiver must at a minimum support the DiSEqC 1.2 standard, which provides control signals for a satellite dish positioning motor. The motorized system will work best if the receiver and motor support the Universal Satellite Automatic Location System (USALS) program.

There is a distinct *disadvantage* to using a motorized dish when more than one FTA receiver is planned. Even though a dual-output LNB can provide signal to two receivers, only one receiver can control the dish's movement, forcing any additional receivers to be subservient to the receiver connected to the motor as far as channel choices go.

Setting Up a DiSEqC-Compliant 4×1 Switch

Several viable combinations are available for using multiple LNBs and dishes. It is not at all uncommon for FTA viewers to set up and use two dishes aimed to receive programming from one or the other of two favorite satellites. One such combination would be the 87 West satellite for its Public Broadcasting Service (PBS) offerings and the 123 West satellite for a variety of programming. The following sections walk through the process for setting up two dishes for receiving any two satellites.

1. Lay in the mounting of the two dish antennas to the proper azimuth, elevation, and LNB offset settings.
2. After the dish settings are made and secured, test/verify that the settings are correct and fine-tune if necessary. To test, first *make sure the power is off to the receiver*, and then connect the cable to the first of the two dishes's LNBs. This brings the relationship to one LNB and one receiver, with the cable going directly from the LNB down to the input jack on a single FTA receiver.
3. Turn on the receiver and perform the setup routine required on your receiver to blind scan the target satellite for available FTA channels. Make sure you set in the L0 frequency appropriate to that LNB on the receiver's setup page. Check a few of the received channels for signal quality and make dish alignment adjustments if necessary to improve the signal.
4. After you are satisfied with the signal quality from the first dish, turn off the receiver power, remove the cable from the LNB on the first dish, and connect it to the second dish's LNB. Turn the receiver on and perform the setup and scan routine on your receiver to collect the available channels from the second satellite while the 1:1 relationship is wired in directly from the LNB to the FTA receiver.
5. Turn off the receiver, and then wire in the 4×1 DiSEqC-compliant switch by bringing a short RG-6 lead from the LNB to the first input connector on the switch. You will probably have to custom build the short leads to length from the LNB to the switch. Use watertight compression-style F-connectors for all the outdoor connections.
6. Do the same to connect the output of the second LNB to the second input on the switch. Finally, connect the output connection on the 4×1 switch to the down-lead to the receiver.
7. At this point, you need to turn on the receiver, and within the antenna setup routine enter the information for switch position for both dishes and target satellites. Most setup routines will move from having DiSEqC 1.1 disabled in the antenna setup routine menu to turning it on, showing the cascade of switches and the port numbers. Set the first dish to cascade 1: port 1 and the second dish to cascade 1: port 2; that should be all that is necessary to set up two LNB/dishes.

The setup menus will vary from receiver to receiver; simply keep in mind that the menu systems on your brand of receiver are essentially toggles and are designed to set up and store associations once you have made the correct selections. The saved setup associates an identified satellite with an antenna connection and then captures and associates a set of channels with that satellite. When you are using the system as a viewer as you change channels from one satellite to the other, the receiver's stored setup information will do the "heavy lifting" and switching for you.

When you are using multiple LNBs canted at a different position on the same dish, the DiSEqC-compliant switch is set up in exactly the same way. Check to verify its signal quality with a 1:1 relationship wired from each LNB to the FTA receiver before connecting to and using the switch port. Checking each LNB/Dish this way will save you frustration and potentially time as you can be confident with each test check that everything is working before inserting the switch into the circuit.

Use a weatherproof or rain-tight enclosure for housing the switches, and put drip loops on all the cables—the cables should go up to the switch and under a rain-tight enclosure. Always use sealed compression connectors on outdoor connections and cap/boot each unused switch connection to keep moisture out.

DiSEqC-Compliant Switch Configurations

Now let's look at what is technically possible. DiSEqC-compliant switches are available in the following Y×Z configurations.

- 2×1 connects two LNBs to one receiver.
- 4×1 connects three or four LNBs to one receiver.
- 8×1 connects five to eight LNBs to one receiver.

DiSEqC-compliant switches are designed and intended only for switching between multiple LNBs, whether the LNBs are set up on one satellite dish or many. Multiswitches, discussed in the next section, are different and are intended to expand the number of receivers that can be supplied with signal from the dish and/or OTA antennas.

Multiswitches and Diplexers

Multiswitches allow you to run cables from a single dish to more than one receiver. The single dish must be equipped with a dual-output LNB, because the various receivers must be able to select between polarizations when alternate voltages are applied to the LNB. Essentially, one of the LNB outputs is tuned to the lower voltage (13V) and the other is tuned to the higher voltage (18V), making all the channels available to any of the connected receivers. The big advantage of a multiswitch when used with a dual LNB is that it can feed signal to as many as four or eight FTA receivers, allowing each receiver to tune in all of the different channels available on the target satellite.

NOTE *Check the switch manufacturer's instructions for maximum length of cable for the down-lead. Line voltage losses can occur at lengths of more than 100 feet, causing the channels requiring the 18V to the LNB to drop out of the mix. When runs of more than 100 feet are necessary, a power inserter can be added to the line to provide the necessary voltage.*

A number of configurations are available in multiswitches. Many models also provide a connection for the OTA antenna's signal to be mixed out to each receiver down-lead. The use of diplexer is necessary on the receiver ends of the down-leads to bring out the OTA signal, as discussed next. Figure 8-1 shows a multiswitch for providing satellite signal to up to four receivers, labeled RX1 to RX4, and OTA to their associated TV sets. At the center top, notice the input jack for an OTA antenna.

Figure 8-2 shows a different multiswitch for up to eight down-leads. This configuration would be useful for larger homes with many TV sets, apartment build-

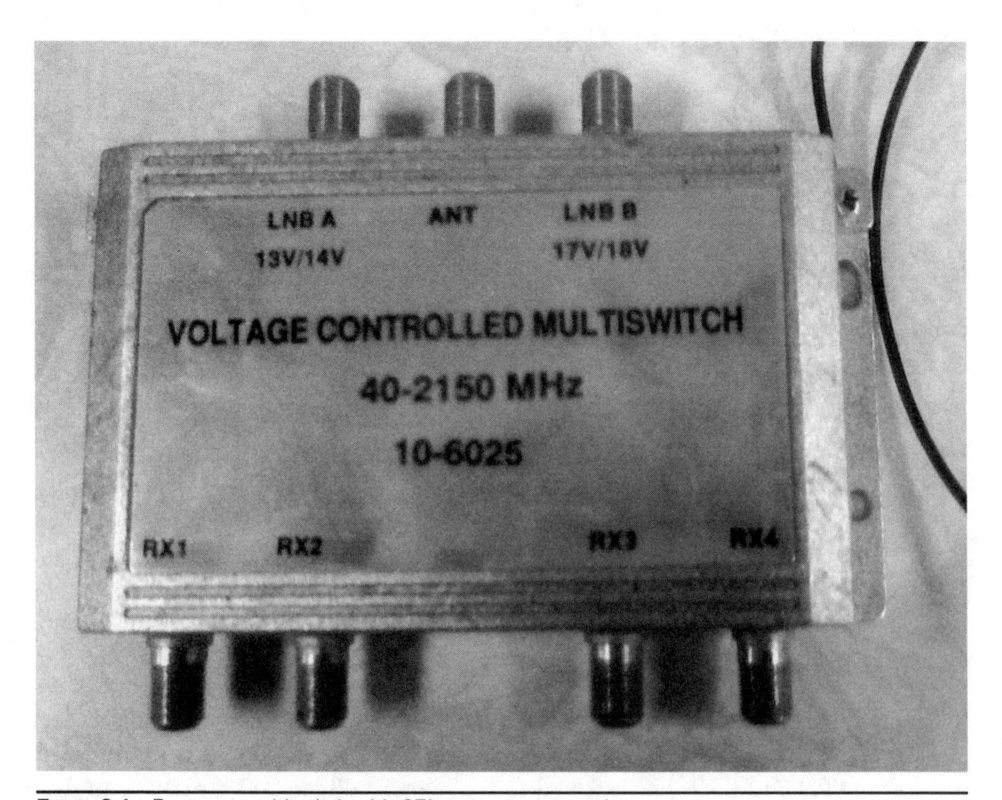

FIGURE 8-1 Four-port multiswitch with OTA antenna connection

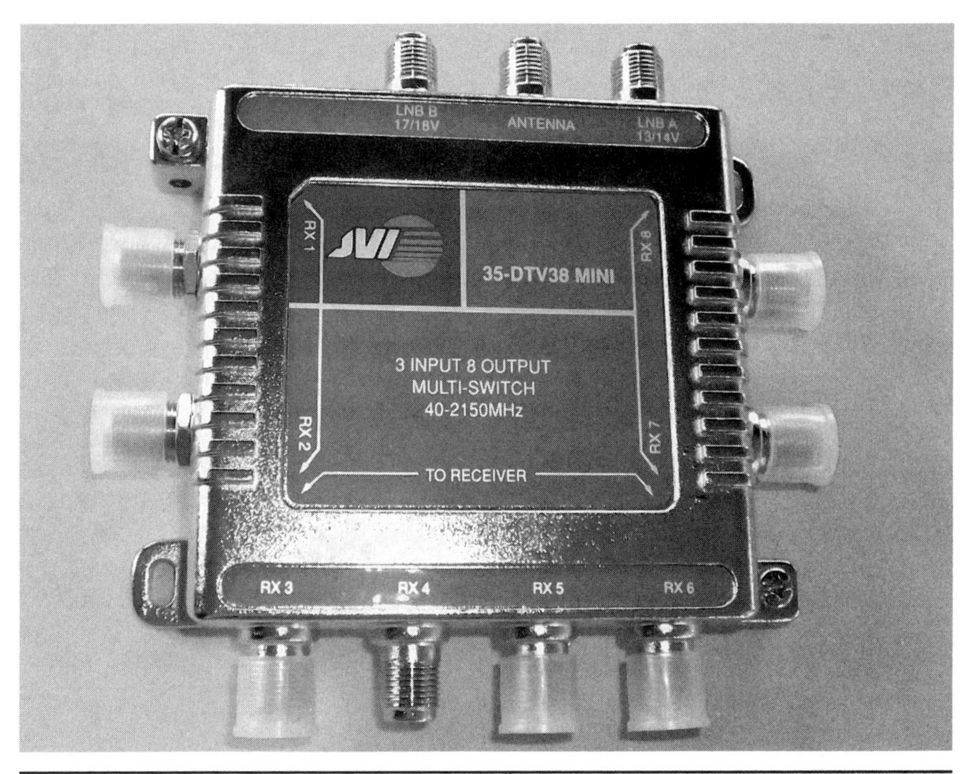

FIGURE 8-2 Eight-port multiswitch with OTA antenna connection

ings, or commercial applications such as offices or restaurants. For most homes, a four-port switch is usually sufficient.

Another cable connection product, a satellite/TV antenna diplexer, is designed to deliver satellite signals and OTA signals down the same down-lead to your equipment. Each diplexer has three connections: a center connection for the down-lead and one connection each for the satellite antenna and the OTA antenna, as shown in Figure 8-3. Two diplexers are needed when used: one to combine the signals outside and another to split them again on the inside of the dwelling. (Radio Shack and Eagle Aspen are both good sources for diplexers.) When a multiswitch is deployed at the dish antenna that is also using the OTA antenna connection, diplexers are used at the TV/FTA receiver end to separate the signals. In some situations, you might want to use this method to bring in signals to both a satellite receiver and a television. As a wiring traditionalist, I think it is less problematic to use separate wiring down-leads for each purpose; however, when cable runs need to be minimized, diplexers will help do that.

All down-leads must be properly grounded using a ground block connector and a 10-gauge or larger diameter copper wire to the building electrical service ground. Figure 8-4 shows a standard grounding block for a single down-lead. The building/

FIGURE 8-3 Standard OTA/satellite diplexer

FIGURE 8-4 Standard grounding block for one down-lead

home/apartment service ground is the location where the electrical service entrance cable is grounded, usually a metal water pipe from the street service or an array of grounding rods.

NOTE *For more information on grounding, check out the National Electrical Code (NEC) articles on antenna and mast grounding. Usually your local library will have a copy of the NEC book in its reference section.*

In addition to grounding the RG-6 down-lead, protection of the satellite receiver can be improved by using a surge suppressor. A dual-lead (two receiver) satellite cable surge protector is supplied by Tii Network Technologies model number Tii 231-2 (www.tiinetworktechnologies.com). A surge protector shuts over voltages to ground to protect sensitive electronic equipment. To be fully effective, surge suppressors must also be properly grounded. Some models of cable surge suppressors rely on the receiver chassis to pick up a ground, and this is not always reliable for providing a minimal resistance path to ground.

NOTE *Each of the connection devices pictured in this chapter has grounding tabs or screw connectors for a ground connection, providing many opportunities for connecting the down-lead shielding to one good ground connection. I prefer to use a ground terminal block at the point just before the cable enters the domicile. Ground connections are necessary for the mounting mast and the cable, meaning two ground wires are necessary. As you think about this, note that the signal wire (RG-6) cable is not connected in any way to the mounting mast, so they both need independent ground wires to carry excessive voltages or lightning strikes to ground and not to your sensitive equipment. Use at least a 10-gauge copper. I prefer stranded bare copper wire for grounding because it is less prone to breaking than solid wire. For that same reason I would never use aluminum wire for grounding.*

Wiring Diagrams for Switches, Multiswitches, and Diplexers

Connecting DiSEqC switches, multiswitches, and diplexers requires bulk RG-6 cable, sealed compression F-connectors, and a few hand tools such as a screwdriver, connector crimp tool, and cable stripper.

Figure 8-5 shows a diagram for a DiSEqC-compliant switch wired with RG-6 connecting four LNBs, to one RG-6 down-lead, to a single FTA receiver. The diagram applies equally to multiple LNBs on a single dish or LNBs from separate dishes. An 8×1 switch can connect eight LNBs to one down-lead, perhaps an uncommon installation, but the components are available.

NOTE *Should you be wondering whether you can combine switches and motorized dishes, the answer is this is not recommended, but if you do this, make sure the motor connection is loaded on the down-lead before the DiSEqC switch. The DiSEqC switch is intended to handle signal switching, not motor current loads.*

In Figure 8-6, wiring from a dish and OTA antenna to two receivers is shown using a multiswitch similar to those shown in Figures 8-1 and 8-2. The diplexer is

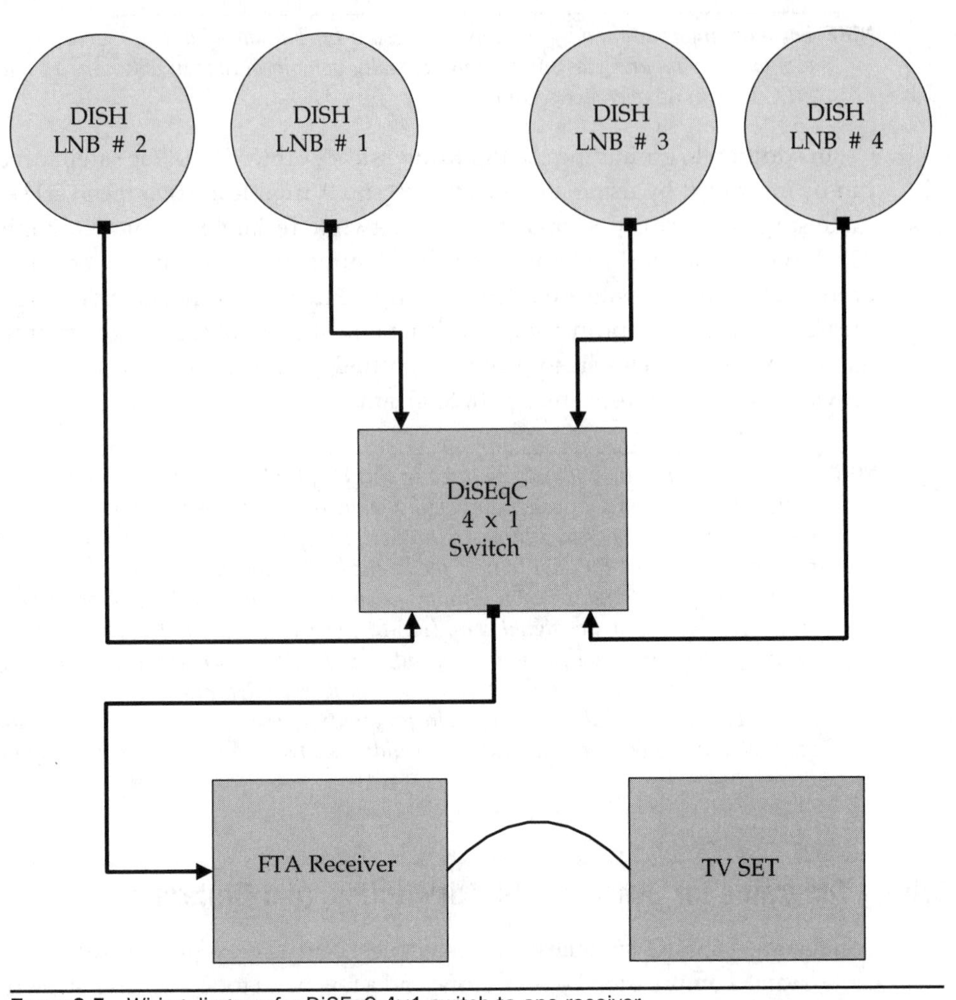

FIGURE 8-5 Wiring diagram for DiSEqC 4×1 switch to one receiver

put in the circuit at the TV/receiver end to separate out the OTA from the satellite. The diplexer also passes the control signals and voltages from the FTA receiver to any LNBs or satellite control motors, without interfering too much with the OTA signals from the antenna. In the diagram, two of the four receiver down-lead ports are not used. Simply cover any ports not currently used with a waterproof boot. The switch shown in Figure 8-2 has plastic boots over the connection ports.

Figure 8-7 shows that two diplexers can be used to bring OTA and one dish LNB signals down to an FTA receiver and a TV set. The cables between the FTA receivers could be HDMI, composite, or component types. The diplexers combine the signals at one end and separate them at the other. When using multiswitches in combination with diplexers, you must follow the wiring labels on the connectors.

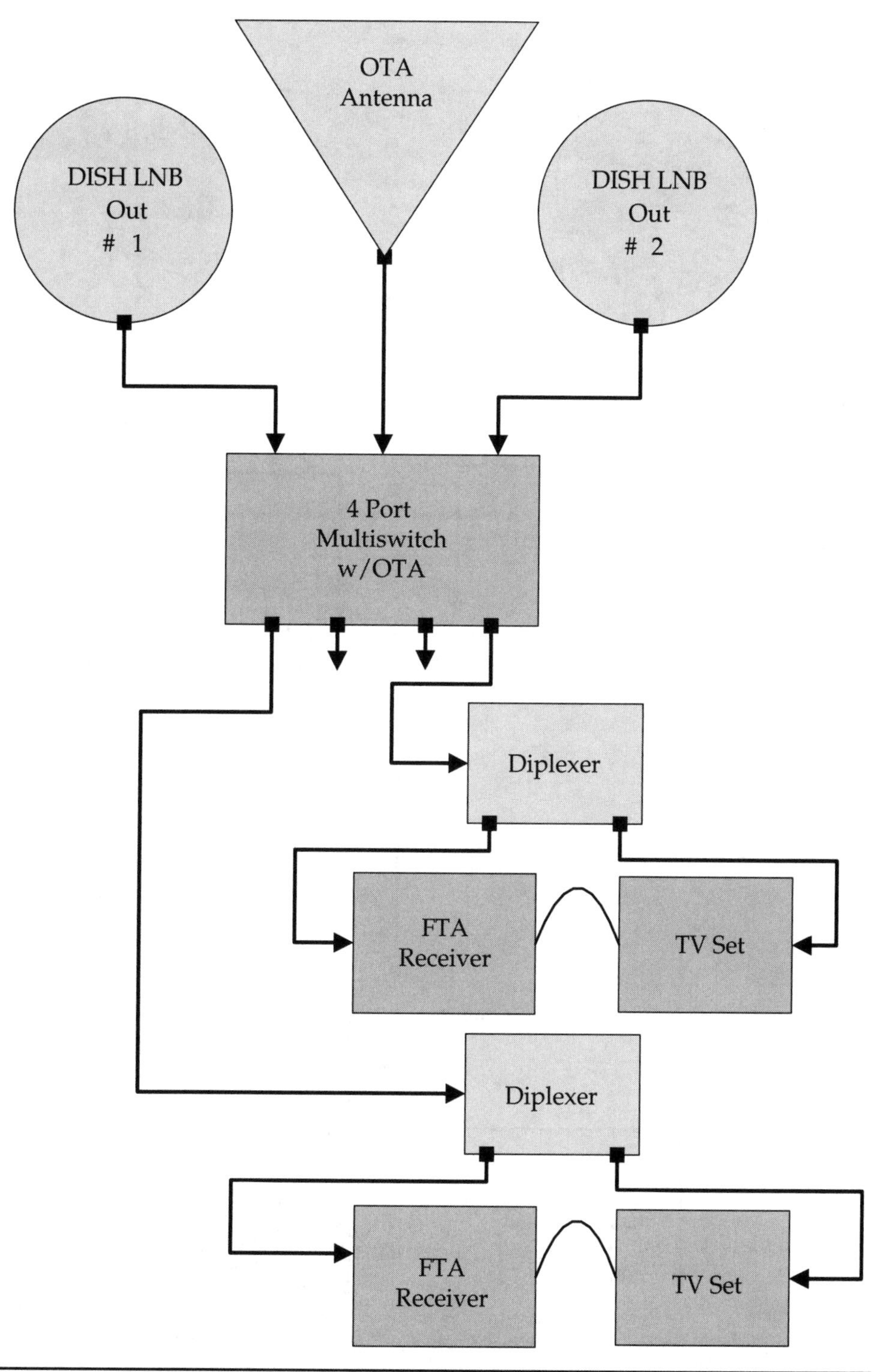

FIGURE 8-6 Wiring diagram for a four-port multiswitch to more than one FTA receiver

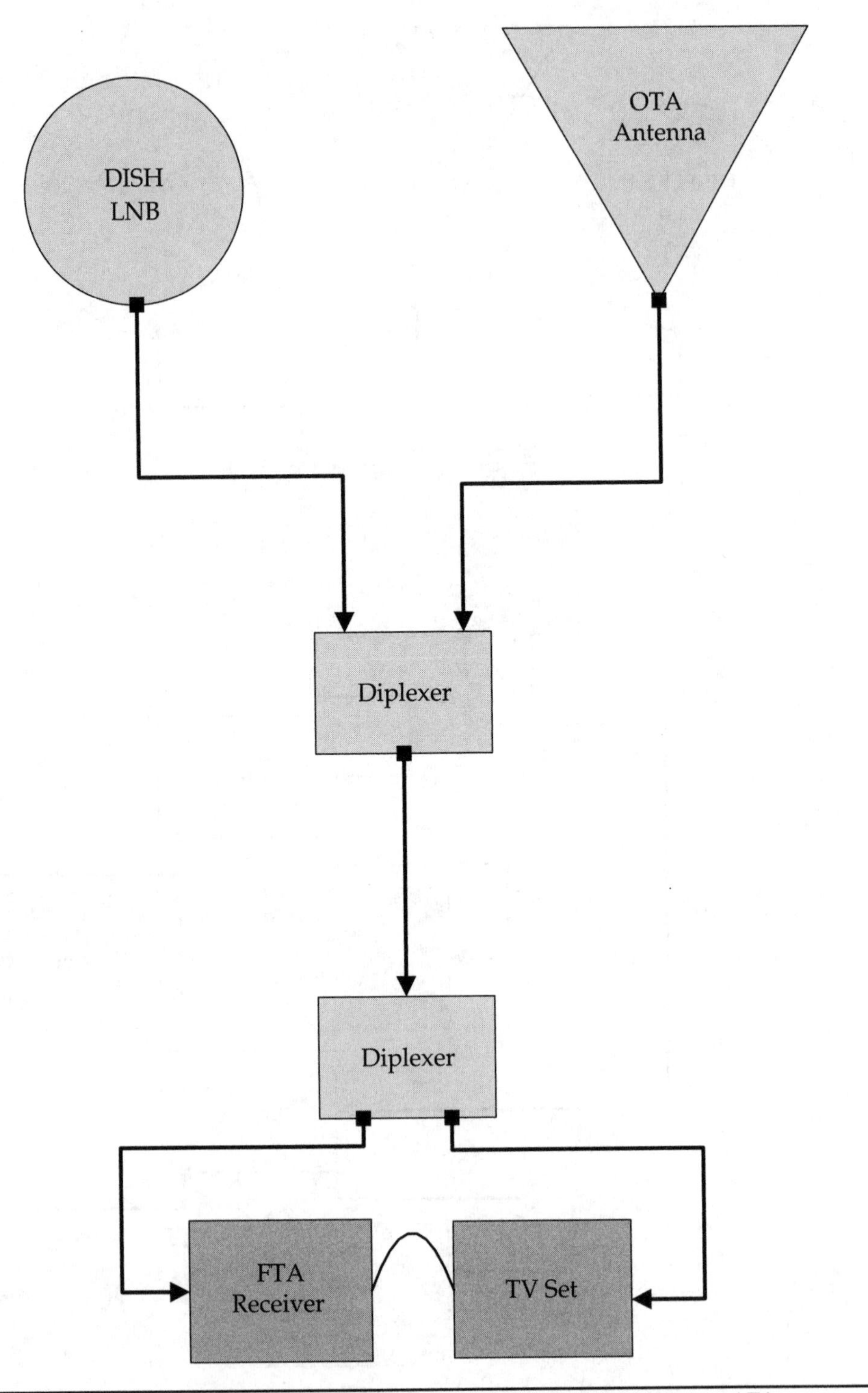

Figure 8-7 Wiring diagram for one dish LNB input and OTA to one FTA receiver and TV

Looking back to Figure 8-3, you can see that the diplexers are clearly labeled for the IN/OUT down-lead and ANT and SAT. At the antenna end, the dish LNB is connected to the SAT and the OTA antenna is connected to ANT. When splitting back out inside the dwelling, the ANT connects to the TV antenna input jack and the SAT is connected to the FTA receiver cable input jack. It is best to use RG-6 with sealed compression F-connectors inside the dwelling as well. Screw-on cable connectors and store-bought slip-on "jumper" connector cables can be problematic, and the cables are often only RG-59 rated.

A Few Last Notes

- To get the most from your single receiver use a single dish with a drive motor, or if using multiple receivers consider using multiple dishes/LNBs. Either choice will contribute to improving the range of viewing options. This can be done right away or over time, as your budget or the FTA muse permits. Decide early on which solution will be best for your circumstances and plan and purchase parts and components accordingly.
- If you are using a pay-for-programming satellite provider, your setup can also deploy multiple satellite dishes and multiswitches providing signals to many domicile receivers. The only difference is the DiSEqC switches and multiswitches will likely be branded with the pay-for-programming provider's logo and will be supported only by their technicians if purchased through them or an authorized dealer.
- Setting up multiple LNBs/dish antennas is not very difficult or expensive, and doing so will enhance the choices available to view with very little added costs. Dish antennas can be purchased for less than $100. Multiple LNB brackets are about $20. Linear LNBs run about $30 each. With FTA, these costs are one-time expenses and will potentially provide many years' worth of viewing choices.

CHAPTER 9

Picking Up Free Local OTA Stations

Prior to the changeover to digital television in the United States, over the air (OTA) television stations were broadcast using an analog radio wave signal. *Analog* simply means that the underlying data such as the sound you heard or the motion pictures you watched were sent though the airwaves by one of two methods: either the frequency of the signal was modulated or the strength of the signal was modulated to carry the data that made up the sound or the picture.

If graphed, the deltas in the frequency or amplitudes would appear as misshapen sine waves proportional to the sound. Similarly, the picture portion would be sinusoidal waveform representing which dots on the screen are to be illuminated. In analog broadcasting using frequency modulation, the frequency of the signal is varied in proportion to the sound of a person's voice or in proportion to a musical instrument. This was not very difficult within bandwidth constraints, because normal hearing for humans is usually within the range of 20 to 20,000 cycles per second. With analog signals using amplitude modulation, the strength of the signal is varied in proportion to an underlying input to carry the picture or sound.

Your car radio likely uses both analog broadcast methods. The FM stations you listen to are using frequency modulation to carry the programming and the AM stations are using amplitude modulation. A graph representing a digital broadcast in contrast would be a varying saw-tooth pattern no matter what the underlying data is. You will notice on your car or home radio that the quality of FM is usually markedly better than AM and much less prone to interference signals from the environment. One of the advantages of the old analog television broadcasting was that the signals would carry long distances over relatively flat terrain.

When the United States switched over to digital television broadcast signals, it essentially changed how the signals were broadcast from analog to digital. Along with that change to digital were sometimes changes in channel number assignments and changes in the strength of the radiated signal allowed from the towers. For digital signals—whether they are carried over the airwaves, over a network wire, a cable TV wire, or a fiber-optic cable—the underlying sound, video, or file is all represented by a string of series of 0's and 1's. A precise time reference is used to determine the beginning or end of a stream of data. Because all that is being transmitted and received is 0's and 1's, it is easy to convey the underlying sound, video, or text file data. With a fiber-optic cable, the light being on for a fraction of a second can represent a 1 and the light being off for a fraction of a second can represent the 0. These 0's and 1's can be used to convey data regardless of the broadcast media and can be at different bit rates within the spectrum allocated to each station frequency.

Depending on where you live, digital OTA television has been boom or a bust. It was a bust for those who were residing at a greater distance from the transmitting tower, where the analog signal was weak but still useable. Many viewers lost the ability to pick up the digital broadcasts entirely even when the same broadcast antenna location is used because the FCC reallocated broadcast power allowance. For those who live close to television broadcast towers, switching to digital proved to be a boom, at least in the number of viewing options, because most channels are broadcasting more than one subchannel within their allowable broadcast frequencies. With analog, if you were watching channel 6 and wanted to change programming, for example, you had to shift to another channel. With digital, the broadcasting station can divide the bandwidth into subchannels. The same old channel 6 might now broadcast channels 6-1, 6-2, 6-3, and more.

The tradeoff for the station and the viewer is the quality level of the video. I often notice as I travel to different parts of the country that TV stations broadcast on subchannel 1 in 720p (HD) and subchannels 2 and 3 in 480p. A modern digital TV or converter box will pick up on the differences of the signal for each channel and preload the subchannels when they are present. So along with enjoying added viewing options, viewers in metropolitan areas also enjoy improved picture and sound quality from the HD broadcasts.

With cable and pay-for-TV over satellite, you will typically receive only the local main channel broadcasts, the dash-one subchannel (6-1, 8-1, 9-1, and so on). Without an OTA antenna, you might be missing out on valuable or entertaining programming carried on the subchannels. This chapter will discuss what you might do to pick up all of the free and available OTA programming in your market area.

OTA, thanks to digital subchannel broadcasting, has substantially increased viewer programming choices in many of the nation's market areas. It is possible that you could be receiving NBC, ABC, CBS, FOX, and CW networks plus Quebo, PBS, and others with only an OTA setup and no cable or pay for satellite bills ever. This chapter's discussion is aimed at assisting you to find out if relying on OTA or supplementing with OTA is a viable option from your location. If your viewing habits are like mine and many people I know, you like to watch programming on a particular network each night of the week. Take the time to find out how many OTA channels are available to supplement your FTA viewing and determine whether cutting the other cords (cable and paid for satellite) that pull money out of your wallet every month will work for you.

OTA Market Area

Broadcast television station licenses are granted in the United States by the Federal Communications Commission (FCC) under rules and regulation enacted by the Congress and signed into law by the president. Two categories of stations exist in the U.S. system: public television and commercial (for profit) television. Public television is supported (paid for) by private grants, some government money, and from donations from viewers who choose to contribute. Commercial television stations are supported by advertising fees. When the FCC grants a license for a channel, the permitted transmitting power is a variable based on that station's market area. The numbers and locations of populations of viewers in the market area that are theoretically served by that station are included in the license application. The viewing areas that are intended to be served are also included in the coverage maps published on the FCC's web site.

Requirements

With modest effort, you should be able to receive all of the OTA channels and subchannels serving the market area where you live, and perhaps with a little extra effort, maybe nearby market areas, too. Purchasing the hardware necessary to receive these free OTA channels will take a little money but nothing in comparison to paying constantly for satellite providers or the local cable companies' bills month after month.

You can potentially increase the OTA stations available on your receiver. Once you are set up, these are free channels to you, broadcast in the clear with no scrambling and for no expense other than a modest amount of electric power to run the electronics. The advertisers for commercial broadcasts are paying the way for you to see these channels and enjoy the programming. As for the public TV stations, usually PBS affiliates, the government, private grantors, and contributors are picking up the tab.

NOTE *Becoming a member of a local PBS station is possible with a small donation, but there is no law requiring you to contribute to watch PBS programming. The same is true for NPR affiliate radio stations; no need for you to pay if you don't want to. That makes all of the OTA TV stations out there entirely free to you if they can be picked up from your location. You can, however, do your part to support nearby NPR and PBS affiliate stations and not part with any of your hard-earned cash in the process. Typically the stations will have fundraiser weeks or weekends, maybe as often as four times a year, where they plead for cash support from their viewers. Most stations need plenty of volunteer help to man phones, to handle mail addressing, and to do other chores during the fundraising weeks. Call the station and volunteer as a way to help maintain this quality viewing option if you can find the time and are so inclined. The choice is yours to contribute if you like.*

Nearby Stations

My first point is obvious, but it has to be stated: One or more nearby stations are required in order for you to receive a signal. So what do I mean by nearby? For digital OTA, nearby is very dependent on terrain. As a general rule, 20 to 35 miles distant from the transmitting antenna over flat terrain would be nearby enough to receive the signal with a modest indoor or small outdoor antenna connected to the TV or to a digital converter box in most homes. On the other hand, antennas in interior apartments in concrete buildings/steel framed buildings might not pick up any stations at all.

When you're examining the FCC web site developed for finding nearby digital TV stations, you'll find that stations are classified by the expected signal strength from any given U.S. ZIP code. The weaker signals will require an antenna with more gain (directional sensitivity). When nearby, medium range, or distant range stations are away from you at many points on the compass, you'll need to have a rotor on the antenna mast or an omni-directional antenna to receive the signals. Omni-directional antennas are a compromise design that will receive fairly well on all points of the compass. A high-gain antenna will receive signals the best when pointed directly toward the station source antenna and has the potential to perform much better than an omni-directional model.

Geography's Role in Reception

A small disclaimer is in order here in the spirit of full disclosure. As you seek to increase your viewing options, geography plays a significant role in what becomes possible for receiving OTA stations from any given location. No equipment will make up for being on the other side of a mountain from a station or being so low in a valley that the signals will not reach. In those cases, your only option might be to pay for cable if it's available or to use FTA satellite if you have a clear view of the southern sky from a location in the Northern Hemisphere. If that same mountain blocks the view of the sky, you can't do much to improve TV reception. TV reception of any kind is a variable that is influenced by terrain, weather, manmade obstacles, electrical and electronic interference, receiver and antenna sensitivity, and station broadcast tower location and signal strength. With luck, your location will fall into the typical situation and will not be adversely influenced by unfavorable localized conditions.

The main concept is this: Is your geographic situation relative to the available OTA stations going to allow you to receive only one or a few stations no matter what you do? Or can you beef up some elements of your installation to increase the channels and programming available to you? When you understand what the possibilities are at your location and have a handle on the cost of, say, a better antenna, adding a rotor, or including an amplifier, you can at that point determine whether the cost of the improvement is worth it to you and your in-home viewers.

DTV-Capable Television

You will need a digital-capable TV to receive digital signals. Any TVs manufactured and sold in the United States after March 1, 2007, must have DTV tuners built in, according to the Advanced Television Systems Committee (ATSC) Standards. Older TVs purchased before 2004 are usually analog and used the National Television System Committee (NTSC) standards. TVs manufactured between 2004 and 2007 can be either analog or digital, so look for Digital Video Interactive (DVI) inputs or HDMI inputs as clues, look at the nameplate data, or hook up to an antenna to see if you can receive the closest OTA station. If you can receive local stations with dash-ones or dash-twos (6-1, 6-2, 8-3) in the on-screen channel report, it is a digital TV.

Alternative: DTV Converter Box

When the only TVs you own are analog and you still want to cut the cord on your cable TV or satellite provider and receive OTA stations for free, you'll need to invest in a digital TV (DTV) converter box, or buy a new TV. The digital boxes run about $40 and typically can be hooked up by rebroadcasting their output on analog channel 3 or 4 to the TV, or you can hook up to composite inputs (red, white, and yellow jacks) on the TV, providing it has them.

OTA Antenna

To get more distant OTA stations, you need some sort of TV antenna. The inputs on any new digital TV are for an F-connector from an RG-6 cable. The old standard, now actually very old, was for flat-wire 300 ohm down-leads from the antenna to the TV. To transition from 75 to 300 ohm, use a matching transformer called a *balun*. (Radio Shack calls it a indoor/outdoor matching transformer: Radio Shack Model 15-1140, Catalog number 15-1140). If you make your own OTA antenna such as a double bay bowtie style, you'll need to use the matching transformer. Chapter 4 discusses some of the basics of antennas and offers some recommendations for you to consider when buying or building an OTA antenna for your TV.

Research

With a digital TV set or converter box and an antenna, your next task is to determine what stations might be close enough to receive—that is, the station tower must be close enough to provide a signal strong enough for your set to process the signal.

The FCC web site provides information about the transition from analog television broadcasting to digital broadcasting.

NOTE *I have no idea how long the FCC web site will stay active on the Web, but the information provided there can be a great help in determining how many OTA stations might be receivable from your home, office, camp, or cottage.*

Here are the questions to explore:

- Which stations (and broadcast tower locations) does the FCC think I can receive from my intended receiving location?
- Which stations does the FCC think are going to have strong signals?
- Which stations will have moderate signals?
- Which stations, if any, will have weak signals?
- Which stations are normally out of reach?
- How many OTA stations do I already receive that might be of value or of interest to me?
- Can changing the ZIP code search by 10 miles or so reveal stations that might be in reach of an antenna rated for any given distance? For example, might a 60-mile-rated UHF antenna possibly be strong enough to receive a distant station of interest?

If saving money and still having the widest range of available programming options is a part of your viewing strategy, OTA combined with FTA is currently the best game in town. Supplement that with a $20-per-month connection to the Internet, and you can enjoy the best of all mass communications methods. So let's

Go Ahead and Use the Antenna

It's OK to be thinking this is a bit weird for older readers who grew up with huge antenna towers fixed to the side of the house with a monstrous antenna affixed to them and rotor control boxes sitting on top of the TV set. Those of us older than 55 watched the towers slowly disappear and give way to cable and huge dish antennas. Sometimes things come full circle and free TV is still free TV, and if that means an outdoor mast model or attic antenna, it is still only a one-time expense. Younger readers may be saying, "Well this is nice, but the future will include free Wi-Fi on every light pole: I'll get my programming that way." That could be true, but it's not likely to happen soon enough. The "Wi-Fi everywhere" concept also disregards the money aspect surrounding available programming. Advertising revenue always chases after the big numbers of viewers. Similarly, content creators want to maximize revenue from the networks and can be reluctant to stream content over a zero or near zero revenue media such as the Internet.

take a look at what's needed to find as many OTA stations as possible in any neighborhood in the country.

Tools and Supplies

A few office supplies are necessary to complete this next task:

- Colored highlighters, various
- Colored pencils: red, orange, green, black, blue, and so on
- Drawing compass
- Internet connection and web browser

Mark Up a Map

Go to the state's department of motor vehicles, transportation, or state police office and find the official transportation department map for your state. If you are 60 to 100 miles *or less* from the border of other states, get their official highway map(s), too. You'll be using colored pencils or markers to mark circles on your map that represent typical ratings of OTA channel 2 through 69 OTA antennas.

1. Mark your TV receiving location with a colored highlighter felt tip pen or colored pencil.
2. Use a pencil compass and the map's legend (scale) to draw the first of four colored circles, with your location at the center of the radius. Mark the first circle centered on your viewing location in black or dark blue at a radius of 30 miles distance. Any TV (or radio) stations within this circle (barring huge terrain issues such as mountains, hills, or valleys) should be within your ability to receive the OTA signal. If TV transmitting towers are located within this area, an inexpensive indoor antenna or modest size outdoor antenna might be all that is necessary.
3. Draw the next circle centered on your location in another color that stretches out to a radius of 40 miles from your location. Station towers between 30 and 40 miles will frequently require an outdoor antenna to be able to receive them. Stations in that zone might be optimal for an omni-directional antenna.
4. Draw a third circle in a different color at a radius of 60 miles. A wide selection of outdoor TV antennas are available from companies such as Winegard and Channel Master that are reputed to receive digital OTA TV stations out to 60 miles. Many of them will be combined (UHF, VHF, FMA/AM) units that can also be the signal source for any quality AM or FM radio receiver. UHF only and VHF only are also readily available. Antennas rated for 60 miles will be bigger, will require mast or tripod mounting, and depending

on your location in relation to multiple stations will likely need a rotor motor on the mast to receive stations from multiple compass directions.

5. Draw the fourth circle in another color at a radius of 100 miles from your base location. The choice field narrows for antennas that are declared functional at 100 miles or more for OTA DTV. Expect to pay a bit more for antennas expected to receive between 60 and 100 miles. The antenna will likely have more elements and more weight, making a very sturdy mounting necessary.

NOTE *Be sure to follow the manufacturer's instructions when mounting any towers, masts, and antennas.*

With the marked maps laid out on a table, you can do some lookups on the FCC's web site. Be prepared to mark up the map with highlighter pens and pencils as the tower locations are disclosed.

ZIP Code Lookup

The database that supports the web site uses ZIP codes to tag the reception data and present it on the web browser. The site will work based on address, city, and state data, but because of syntax issues, the easiest method is to know the ZIP code of your base.

1. If you have an Internet connection already and a web browser, you can use the U.S. Postal Service's web site to look up ZIP codes (http://ZIP4.usps.com/ZIP4/citytown.jsp). Enter the name of your own viewing location in the ZIP lookup, as shown in Figure 9-1, which shows the entry fields of the ZIP code lookup site with an entry for a small village in Michigan's Upper Peninsula.

FIGURE 9-1 ZIP + 4 "by city" lookup entry fields

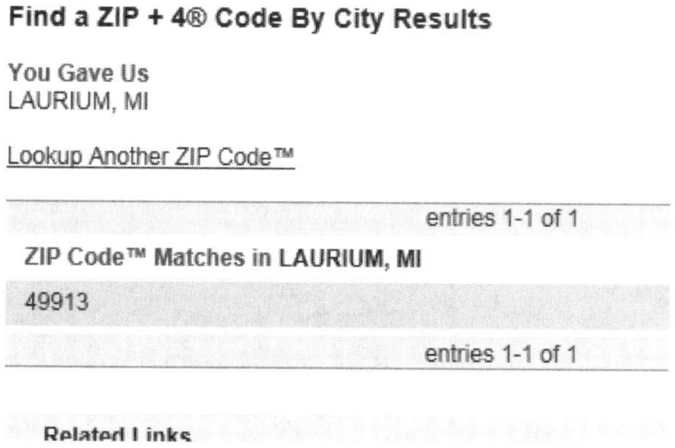

Find a ZIP + 4® Code By City Results

You Gave Us
LAURIUM, MI

Lookup Another ZIP Code™

entries 1-1 of 1

ZIP Code™ Matches in LAURIUM, MI

49913

entries 1-1 of 1

Related Links

FIGURE 9-2 ZIP code data for Laurium, Michigan

2. Click the Submit button, and the database returns all of the ZIP codes for that city, village, or town, as shown in Figure 9-2.

NOTE *Another way to find the ZIP codes is to look in area phone books or take a trip to the local post office to use the ZIP code directory book there. Or hike, bike, or drive to the local library and find the reference desk or use a computer there.*

DTV.GOV

For the next steps, an Internet connection is necessary. You will begin the search with the ZIP code of your TV viewing location and then work your way out within the 30, 40, 60, and 100 mile circles to neighboring towns and cities. TV station locations and coverage areas are somewhat market/population driven, so use the bigger nearby population centers for the searches. Here we'll examine some representative examples from various areas around the country.

At home, at an Internet café, or at a public library, with ZIP codes and your maps in hand, bring up the government's DTV web site at www.fcc.gov/mb/engineering/maps/. This site is used for looking up the FCC engineering data maps and station IDs with their interpretation of the various station's signal strength in a given ZIP code area. At the upper left, you'll see the Enter Location field for entering the ZIP code, as shown in Figure 9-3.

Click the Go! button and the database will return a Google Map with the ZIP location showing on a map at the right and a column on the left listing the station call IDs. Use these to look up the FCC engineering data maps and station information and the FCC's interpretation of the various stations' signal strengths in that ZIP code area. The black and white photos in this text will not show it well, but the signal strength ratings are color-coded. Four green bars indicate "strong" signals,

DTV.GOV WHAT YOU NE
HOME LEARN ABOUT DTV DT
Enter Location:
Go!
Address, zip code, city, etc
DTV Reception Map Tips
If searching on city, include state
If searching on street address, include city and state
Exclude PO Box, Mailbox, APT # in address search parameters
Avoid using special characters, such as @, #, $, %, *, etc.

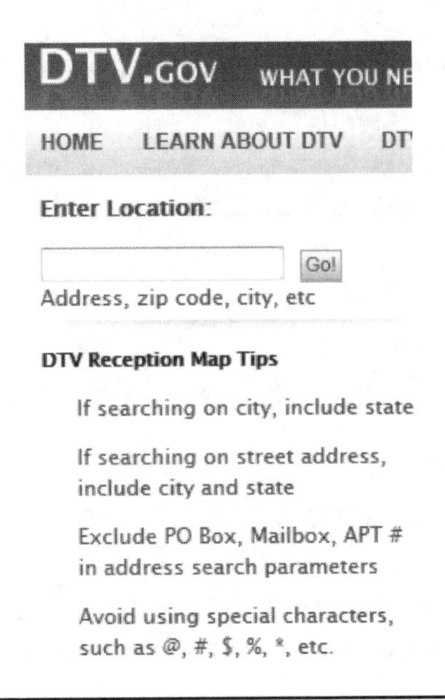

FIGURE 9-3 Entry field for ZIP code

three yellow bars is "moderate," two light orange bars are "weak," and a pink box with red X is for insufficient signal strength. Take a close look at Figure 9-4 to see the station IDs. Notice that the relative bands of the stations are called out of the database as well such as Hi-V (HF) for WNMU.

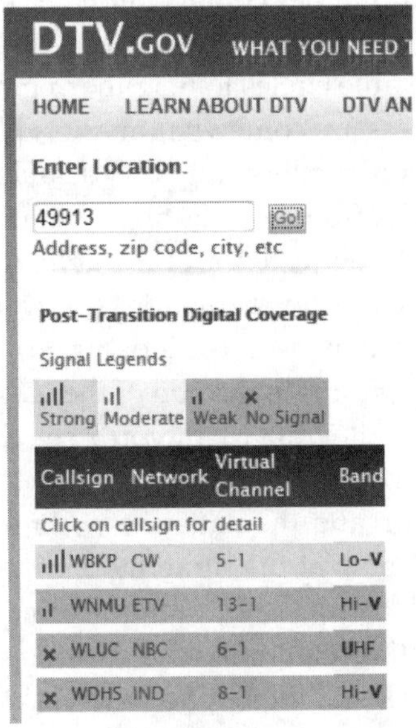
DTV.GOV WHAT YOU NEED
HOME LEARN ABOUT DTV DTV AN
Enter Location:
49913
Go!
Address, zip code, city, etc
Post-Transition Digital Coverage
Signal Legends
Strong Moderate Weak No Signal
Callsign Network Virtual Channel Band
Click on callsign for detail
WBKP CW 5-1 Lo-V
WNMU ETV 13-1 Hi-V
WLUC NBC 6-1 UHF
WDHS IND 8-1 Hi-V

FIGURE 9-4 Data returned from FCC database based on ZIP code 49913

Figure 9-4 tells us that WBKP is a CW network station, it is broadcast on OTA channel 5-1, it is in the Lo-V band, and the signal strength is strong from this ZIP code. This is useful and interesting data in itself. The next station on the list is WNMU, an educational TV station with a weak signal in the ZIP area.

More interesting information is what you can see when you click the WNMU station's call sign. Two things happen at that point. First, the left panel of the web page expands to show the information shown in Figure 9-5: azimuth and relative power. The magnetic azimuth (indicated in text) to the station tower and the Google map to the right of the page displays the actual compass azimuth line to the tower.

Click a second station on the list, and the same two responses are produced. First the data in Figure 9-6 is presented, and then the Google map leaves the first station's azimuth and adds the second. In this case, they are separated by 63 degrees, indicating that a rotor might be necessary to swing an antenna to pick up one or the other station. A closer look at the Google map expanding the area on the map shows that channel 5-1 might be close enough to receive, even if the antenna is pointed at the more distant channel 13-1.

At this point, when you are doing your own area, you would expand the Google map sufficiently on the web page using the plus and minus buttons to disclose the relative tower locations and mark them on your paper map. Then you'd draw in the azimuth line to your location on your map.

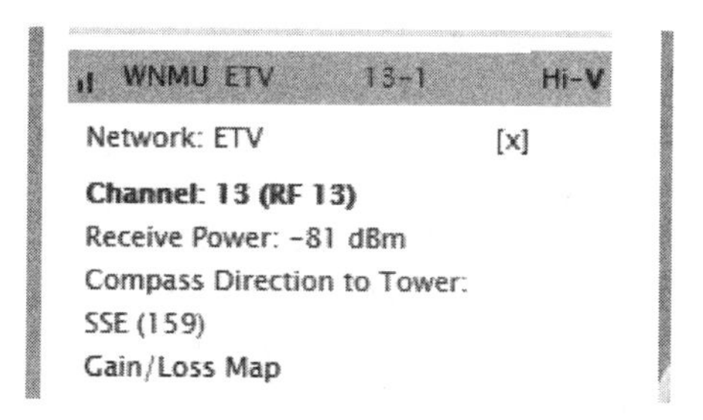

FIGURE 9-5 Channel, power in dBm, and azimuth from ZIP 49913 to the tower for channel 13

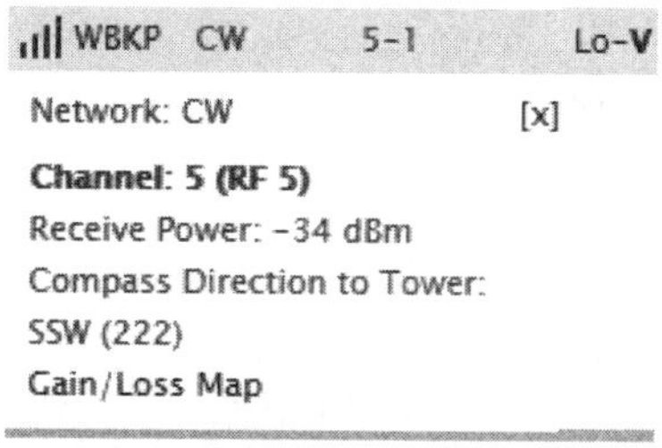

WBKP CW 5-1 Lo-V
Network: CW [x]
Channel: 5 (RF 5)
Receive Power: -34 dBm
Compass Direction to Tower:
SSW (222)
Gain/Loss Map

FIGURE 9-6 Channel, power, and azimuth from ZIP 49913 to the tower for channel 5

TIP *It will be helpful later if along the azimuth line you mark in the station's ID by stopping the azimuth line and writing in the call letters and then continuing the line (for example, --------WKAR-------).*

Regarding the dBm numbers displayed on this page for signal strengths as a decibel expression of radio wave power—the dBm is a log function; in this case in comparison to 1 mW or 0.001 watts, or alternatively it can be expressed as 0dBm. Comparatively, 100 mW is expressed as 20 dBm. The data on this web site is relative signal strength, not receiver sensitivity. The negative values stated for dBm means that they are log values below 1 mW (one milliwatt). For example, one hundredth of one milliwatt of power would be expressed as –20 dBm. With 0 dBm as our reference point, keep in mind that the closer the negative dBm numbers are to zero the stronger the signal is in that ZIP code. In our examples Figures 9-5 and 9-6, the respective signal strengths are –81 dBm for WNMU, –34 dBm for WBKP, and so we know both are under 1 mW. WBKP has a stronger signal at this location by a significant factor. To keep the comparison in round numbers, the –34 value is close to 0.001 mW and the –81 value is close to 0.00000001 mW. Not to bore with the math, but it is easier to see that the –34 dBm is a stronger signal than when comparing watts to watts. The WBKP tower is providing a much stronger signal at that location because it is closer to becoming a positive number. The reverse is true when looking at a receiver's sensitivity levels: the larger the negative number, meaning the farther it is from zero, the weaker the signal it can successfully deal with and give you audio and video from it. Signal strength and sensitivity are inverse relationships.

While still on this web page, you can do one more thing—click the call sign for each station to reveal a highlighted link for the Gain/Loss Map. After you have the TV station towers and azimuth marked on your map, sketch in the "footprint" of the radiated signal of the OTA station using a colored pencil. Use the same technique as above to identify the station footprint outline by its call letters ---WKAR---, and use a new color for each one.

One such footprint map is shown in Figure 9-7. This map page shows the delta in coverage when the station was switched to a digital license and the change in radiated power. You will want to trace out the solid lines (digital coverage) on your map for station coverage. The dotted line information is for analog and is now obsolete.

NOTE *This example also brings to light an interesting feature of digital TV. Many TV stations considered the station channel number as a brand or trademark and wanted to keep them. A digital receiver or converter box displays this particular station as channel 6-1 for the viewers, even though the station is actually broadcasting on channel 35, as shown at the top of the page of Figure 9-7. This feature on digital TVs is called "channel virtualiza-*

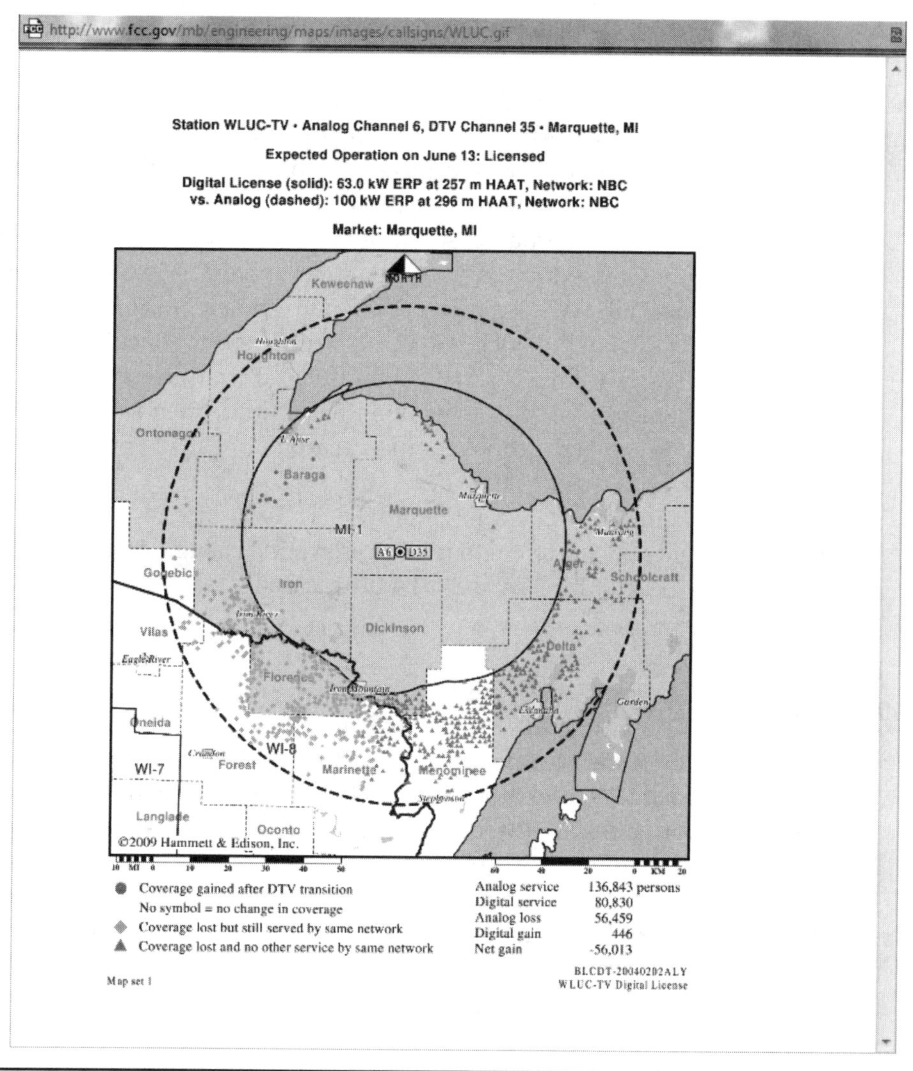

FIGURE 9-7 Expected coverage area for WLUC-TV 6

tion." Keep that "feature" in mind when you're considering antenna selection, particularly if you plan to buy a Yagi style antenna. It's nice to know the real frequency you are trying to receive, particularly when you are just outside of the expected receiving zone (footprint). There is a big difference between UHF and VHF antennas, as discussed in Chapter 4.

The HAAT number is decoded as the center height of the transmitting antenna in meters above the average terrain measured above sea level radiating out from the tower at 45-degree intervals from 3 to 16 kilometers away from the tower. The

ERP numbers are for effective radiated power in kilowatts at the transmitting antenna.

Trace out the coverage area for all the TV stations from your ZIP code.

Next, to see if you can expand the viewing options, pick some higher population center's ZIP code about 10 to 40 miles away from your location in all directions to see if the numbers of potentially available channels improve. Keep in mind that it is difficult to receive stations past 100 miles away, even with a high quality antenna and rotor. Going out to find towers much beyond 100 miles will usually not be worth your while. Where a network affiliate is noted on the FCC license info, such as NBC or FOX, it might not be beneficial to go through a lot of effort to "bring in" the same network from a distant station if you already receive that network. There are times when it helps to have two PBS stations, such as when local programming such as the high school chorus's seasonal regalia is broadcast in your favorite network show's time slot.

One more example is shown in Figure 9-8 for Twin Bridges, Montana. In this example, the highest signal strength is weak. Elevation changes in parts of this state make reception difficult for FTA and OTA broadcasts.

Choose the Best Options

With the sketches of the coverage maps transferred and scaled to your state map, you need to do some terrain and front/back yard analysis. Study your situation and use the information here and from other sources, and decide what the best options are for your unique circumstances.

Knowledge of local and nearby geographic conditions is necessary so you can analyze what is on the working map and known to be on the ground to make an

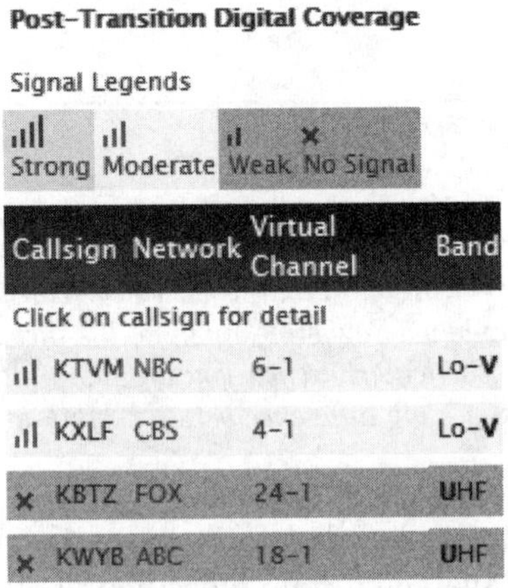

Post-Transition Digital Coverage

Signal Legends

Strong Moderate Weak No Signal

Callsign	Network	Virtual Channel	Band
Click on callsign for detail			
KTVM	NBC	6-1	Lo-V
KXLF	CBS	4-1	Lo-V
KBTZ	FOX	24-1	UHF
KWYB	ABC	18-1	UHF

FIGURE 9-8 Twin Bridges, Montana, station availability

informed estimate of the stations you should be able to receive and what equipment might be necessary. The equipment list worst case is to upgrade the TV or converter box to one with a highly sensitive receiver. Add to the list a 100+ mile antenna, an in-line preamplifier, tower, and rotor.

If your location is in a sheltered valley surrounded on all sides by mountains, nothing you can do will allow you to receive OTA. In all probability, the situation will be somewhere between the extremes, and your location and viewing opportunities will be significant; the only added expense might be for a better antenna. Keep in mind that it is possible to be land-locked out of a signal's path in a city environment as well. Being surrounded by high rises does not help much either.

RF Line Preamplifiers and DVRs

You might want to add two more devices to your OTA system: an in-line amplifier between your antenna and TV receiver or converter box and a digital video recorder (DVR).

In-line preamplifiers are available from a number of manufacturers for less than $100 and will amplify channel 2–69 signals from your antenna. One such unit that allows combined or separate input antennas for VHF and UHF is the Channel Master CM 7777 Titan2 VHF/UHF Preamplifier with Power Supply (model CM7777). Preamplifiers are also helpful if you run more than one TV off of your antenna or have a long cable run from antenna to the TV sets.

If you want to change the time gap for watching your favorite shows, you can do it even without cable providers or paid-for satellite digital boxes. Some viewers still use a VCR to record a show to watch later. DVR is also possible now in conjunction with OTA programming. Channel Master has a vested interest in increasing the numbers of OTA viewers for obvious reasons.

One long-held objection to switching away from paid services to OTA is that you can't record *The Price Is Right* at 11 A.M. and watch it at 7 P.M. on the cable box. But Channel Master has a solution for that problem for OTA viewers with its entry of an OTA antenna–compatible DVR. The Channel Master CM-7000PAL Antenna Compatible DVR (CM7000PAL) might be available in a store near you. It's worth looking into if you want to warp OTA TV time and you have about $300 to spend.

For readers who do not want to give up basic cable or their paid-for satellite plan, you can easily add local OTA reception to any television set with multiple inputs. Perhaps the cable box is taking up the RF (antenna) input on the TV. In this case, a separate digital TV converter box can be added to one of the composite input jacks on the TV and the OTA antenna can be connected to the digital converter box. If you want to view OTA programming and channels, you can switch over the TV's display/input to the digital converter box.

CHAPTER 10

Selecting the Optimal TV Simplified

The feature choices out there now for buying a new TV are both incredibly amazing and sometimes daunting for many consumers. Too many sets, too many features, and way too much deciding are involved as you try to sort through the hype taking up ink in the Sunday paper's advertising flyers. Computers are working as TVs, TVs and DVDs are now browsing the Internet, and it's possible with some TVs to make phone calls over Skype! The days where selecting the finish color of the faux-wood case might have been the most important decision are long gone. Televisions are slowly becoming, and maybe already are, communication devices that make up part of a communications and entertainment system. Today's TVs do more than simply present the sound and motion pictures from a radio wave signal from a local TV station. Fortunately, they still do pick up local stations, and they do it very well.

Choosing TV Features

As you consider buying a new TV, it pays to take a few minutes, or a few hours, to consider what features are most important to you and how much you will reasonably have to spend to get those features. If you are buying only one television, you will probably, as the clichés go, want to get the most bang for the buck, the most thunder for the plunder, the highest value for each dollar spent. But buying a TV with all the currently available features can be expensive.

Have your important features list with you or in mind as you browse the stores to find out what is available. I think the most important advice of all is to take a look at TVs in the store while they are on and working. Ask salespeople to let you see the TVs receiving a channel you typically watch and see them playing a DVD. As you look at the different screen technologies and brands and compare one to the other, let your eye help you decide. When the set is running in your home, you are the final judge of picture quality. Most people tend to use a television for many years, so it is nearly as important as choosing a car. It's a choice you are likely to live with for a long time.

Quantity

If you are in the market for more than one TV, consider where and specifically how each will be used. The location of use will play a major influence on what features will be important to you and the viewers in each location. Look for a set with only the most important features to help you save money and avoid paying for features that will never be used.

Perhaps you enjoy listening to the evening news while working on projects in the garage after dinner. Having a TV in the garage that will connect to the Internet or play DVDs might not be important to you, so the simplest set with the fewest connections and features might be completely sufficient for this location. The simple TV might also be perfect for a small kitchen unit that gets viewed only while you're making dinner or cleaning up afterward.

Consider these areas where television viewing is typically of a short duration and the secondary purpose for the space as casual TV viewing areas. If your family is like most families, big or small, a few areas will become or are planned as primary TV viewing areas—the living room, family room, a multipurpose/game room, or a home theater space. Many people also enjoy stretching out and watching television or movies while lying in bed, so one or more bedrooms can become important secondary TV viewing areas. Sorting the spaces in this fashion allows you to build a most important features list with the group of features that are imperative for that space and how each television set and its accessory equipment will be used and enjoyed by viewers in that room.

Budget

Budget might be your only important factor. If you are buying on a limited or fixed budget, TV feature options can be limited by list price or available sale pricing. Even when buying to a budget, you can still develop and use your "most important features list" before you look for sales and bargains. Equipping yourself with this list will help you stay within or even under budget and avoid succumbing to the up-sell pitches of high-pressure commissioned salespeople at the big box store. Knowing exactly what size and features you need and what you are willing to pay will simplify the purchasing process.

Screen Size and Display

Over the air (OTA) consumer televisions are readily available with screen sizes from 2.4 inches to 150 inches. Some very low production volume television units are offered at less than 2 inches and more than 205 inches, both of which are very disproportionate for the average living space. Select a screen size appropriate to the viewing area where the television will be used. If your budget prevents buying a large screen for a large room, that might limit the seating options right away—sitting too far away from a 12-inch TV, for example, can be particularly frustrating when you need to read or view fine details on the screen. Having to sit too close to a television that is very large can prove equally frustrating and detract from the quality of the experience.

Being able to read the on-screen channel display and menu prompts without getting up from your chair and moving closer to the TV is a critical feature. Relaxing TV viewing and exercise do not mix well. At my home, without wearing glasses I can read the on-screen channel display on my Sansui 19-inch TV out to 9 feet from the screen. When I get past that mark, it becomes difficult, and it is impossible for me to read the display at 11 feet from the screen. This readability factor is not listed in the technical specs for any television; you need to test it out in the store before buying.

A relationship exists between screen size and minimum, maximum, and optimum viewing distance from the set. The viewing rule of thumb that says an adequate viewing area equals three times the width of the screen is less than perfect in my experience, simply because as screen size increases, the range distance of acceptable and comfortable viewing becomes larger. Having said this, the rule is not perfect for everyone's individual vision and viewing preferences. As you test your personal preferences, you will likely find an optimum television viewing area at a distance of somewhere between two to five times the diagonally measured screen size. For example, with a 42-inch screen size, the seating should not be closer than about 82 inches (6.5 feet) or much farther away than 210 inches (17.5 feet). With screen sizes larger than 32 inches, the maximum viewing distance might work out and still be OK at six times the screen size, but it is not a good distance for extended viewing or best quality viewing.

Flat Panel Display

Flat panel TVs and displays are those without curved glass surfaces and tend to be of minimal depth. Flat panel TVs are now the norm, not the exception.

LCD Display

This type of display, monitor, or TV uses electrically charged liquid crystals to emit the light making up the reproduced picture or image. LCD lights are also common in places such as car dashboards and flashlights.

Plasma Display

This type of display, monitor, or TV uses electrically charged cells of trapped gasses to emit the light making the reproduced picture or image. Plasma screens are typically manufactured in sizes that exceed 30 inches. Fluorescent tubes or bulbs emit light from electrically charged "plasma," so the technology itself is not new, but it is relatively new to flat panel television displays.

Viewing Angles

In a TV's technical specs, you'll see a rating for viewing angle expressed in degrees. Figure 10-1 shows you an overhead view of what that means for a horizontal viewing area. The figure represents a set with a 32-inch screen and a generous clear viewing angle of 160 degrees. If someone is trying to see the screen outside of the available viewing angle, the picture can range from nonexistent, to extremely dim, to having way too much distortion. Sitting at the extreme 160-degree angle would not be great either.

Consider the viewing angles necessary to accommodate viewers in each room that will have a TV set. A larger viewing angle allows increasing the available "front row" seating capacity. Keep in mind that viewing angle limitations can exist in both the horizontal and vertical planes. An LCD set with a limited viewing angle in the vertical plane might not be the best choice, for example, for mounting high on a wall in an office waiting room. Screens can be designed with a tilt or built-in bias

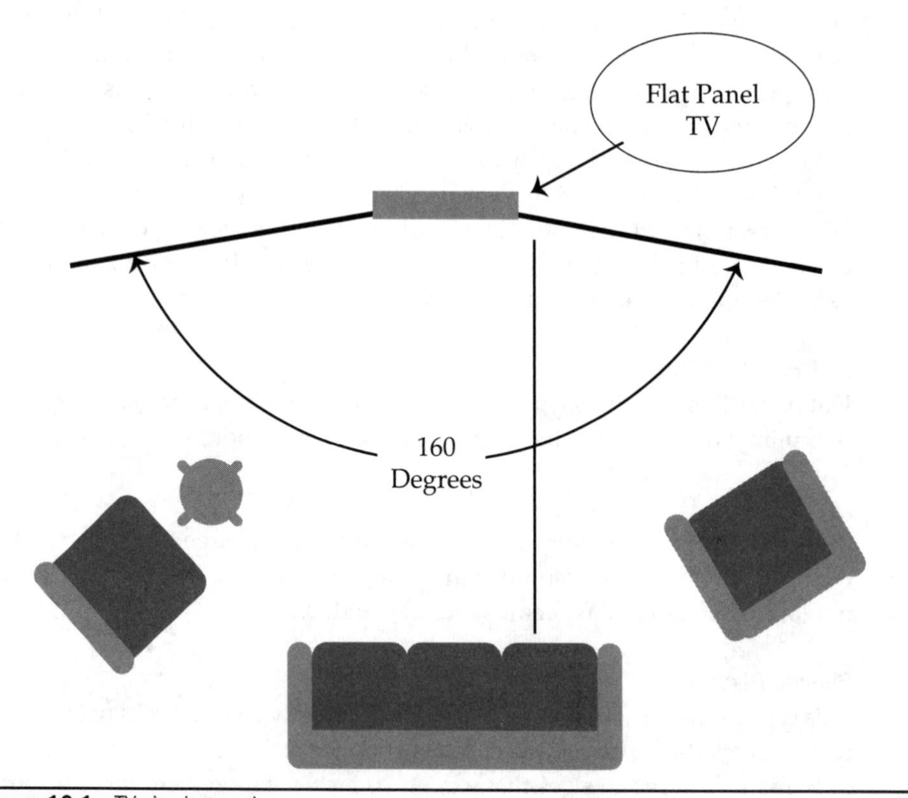

Figure 10-1 TV viewing angle

for viewing from high or low mounting if needed for a given application. Also with high wall mounting locations, you can tilt the whole set's mounting on a downward angle to improve viewing quality for the audience.

Viewing Locations

Optimum TV viewing locations for the primary, secondary, and casual viewing areas in your home or office are determined by a combination of factors that include the available space, the relationship of screen size to viewing distance, and the acceptable viewing angles defined in the TV's specifications. I have learned that finding potential mistakes is much less expensive if you discover them on paper ahead of time instead of after you purchase the TV. Based on that, I recommend that you measure and sketch out TV locations and viewing areas before you make a purchase. Scale the distances and use a protractor for the angles. By doing this, when you go to the store to buy a TV of a given size for a particular location, you can "test view" it in the store at the calculated distances and angles.

Screen Resolution

The visual quality of a TV screen or display monitor is primarily composed of two independent factors related to the display's clarity. The first is the size of the display pixels that emit the light making up the video picture. This size is reflected in a measurement related to the number of pixels per unit of distance. For example, a screen could have 10 pixels per inch on the horizontal plane and an equal or different number in the vertical plane. The second factor is the count of pixels that make up the display in the horizontal and vertical. The two most common pixel counts in currently marketed televisions are stated as 720p and 1080p. These numbers refer to screens or displays where the pixel count is 1280 horizontally by 720 vertically or one that displays 1920 horizontally by 1080 vertically. The advantages of having these higher resolutions are covered in more detail in Chapters 6 and 15.

Refresh Rate

Refresh rate refers to how many times per second the pixels are re-energized to emit the light you see in the picture or video. Typical refresh rates are 60 Hz, 120 Hz, and 240 Hz (Hertz). 1 Hertz equals one cycle per second. The significance of refresh rates are related to how motion video is reproduced. Lower refresh rates tend to blur fast-moving objects in the video and faster refresh rates tend to represent motion in the video with more clarity. When "talkies" are filmed with a 35-mm motion camera, the pictures are taken and replayed at a rate of 24 frames (pictures) per second. Because filming, even with digital equipment, is taken at or near 24 frames per second, the higher refresh rates do not provide exponential improvements in the reproduction of motion, but the difference is enough to notice.

Green Factor

Any TV or piece of accessory equipment specification will provide the power consumption in watts. A fair way to compare the relevant "greenness" of one television over another is to compare based on watts per square inch of viewing area. The least watts consumed would have the lowest environmental impact and have a lower cost of operation for a given size television. Typically accessory devices such as DVD players with similar features are usually very close in their wattage consumed when in operation. Simply compare the label information. If maintaining a minimal environmental impact and lowering your cost of operation is important to you, include comparisons of power consumption for the TVs you are considering for purchase. TV sets are currently available that use very little electrical power to run. One example I found at a discount chain in North Carolina was a 22-inch set that used less than 70 watts.

3D and 3D Glasses

Yes, you can view 3D movies at home. You will need a television display capable of decoding and displaying the video stream for 3D and a pair of 3D glasses for viewing the image. Pricing on decent televisions using this technology tend to start at about $1500 and can easily reach $5000. If you watch the same 3D movies on an LCD set and see them again on a plasma model, you will likely appreciate the image quality improvements present in a larger plasma set.

Available I/O Connection Types

Televisions circa 1958 had one input, two screws on a piece of Bakelite on the back of the TV chassis, to connect OTA analog radio frequency from the 300 ohm two-wire down-lead with the distant end connected to an outdoor antenna or from a "rabbit ears" indoor antenna. That was its only input. In my neighborhood, that meant receiving only two channels: Channel 6 from Marquette, Michigan, and Channel 2 from Thunder Bay, Canada, across Lake Superior. If you aimed the antenna just right, you could pick up both stations well without a rotor on the mast.

We could long for the good old days when things were much simpler, but simple also often means limiting and doing without many fun features and capabilities. Instead, let's attempt to make some sense of the connection options that are available today and perhaps a little less complicated for the non-guru. At this point, you might be thinking "Why does this matter?" Well to start with, there has to be a way to connect the video and audio output of the FTA satellite receiver you are hooking up to your TV as a part of this project. Also you will likely want to connect up one or more DVD or Blu-ray disc players as well.

Using the various inputs on a TV to bring the video to the screen and audio to the sound speakers from a secondary source is a simple puzzle in which you make a four-way match. This four-way match involves the available outputs from the player or device, the input selection switch or menu on the television, and the input/output jacks on the TV and device. If you can mix up a latte (water + ground espresso coffee + steamed milk + favorite syrup) or a batch of chocolate chip cookies, you can easily make the necessary device matchups to the TV. It is very much like following a recipe.

First, the source device must have an output connection or jack suitable to an input jack on the television set. Next, you have to have the correct cable or cable set to connect from each input jack to each output jack. Sounds logical so far, right? Not much more difficult than plugging a lamp cord into a wall outlet. Where many people have difficulty is in the final necessary step in the process. The third step is basically a matter of using the TV's menu selection, via the remote control, for input choices, or using the input (sometimes labeled "device") button to toggle through to the correct selection to match the input jack to which the source device (DVD for example) is connected. Easy as pie on paper, and easy as pie in the real world, too, with just a little bit of patience and practice. The biggest fear many people have is that something is going to be damaged permanently by using the menu or changing the input button. Be assured that these interfaces are there to use, and anything you might "mess up" can be "un-messed up" with another push or two on the button or remote. It is no different from changing the heat settings on the stove or shifting gears in the car. The choices are programmed into the devices for your use and, short of using some inappropriate hammer throwing, you can't hurt the TV or devices by switching things up a bit.

The I/O (input/output) connections on an example small flat screen television are shown in Figure 10-2. Starting from the left side of the figure, the I/O connections for this TV will be discussed in the next few paragraphs. This detailed example and discussion should help regarding the inputs when you're selecting a new television, and surely should help when you're connecting your own, as the saying goes, "insert tab A to slot B." For the TV sets and players that you already have, you might want to take a look at the connections available on them as you follow along with this text to determine what connections options your current TV and video equipment has available. Figure 10-2 is an example of what might be on a TV you already have in your home or one you might purchase. Take a good look at as the figure, because the text will frequently refer to it as the various interfaces are discussed.

Component Connector Jacks

At the far left in Figure 10-2 are five RCA connection jacks for connecting a component—usually a DVD player. Although impossible to discern in the black and white format used in this book, the RCA jacks are color-coded. They are also grouped

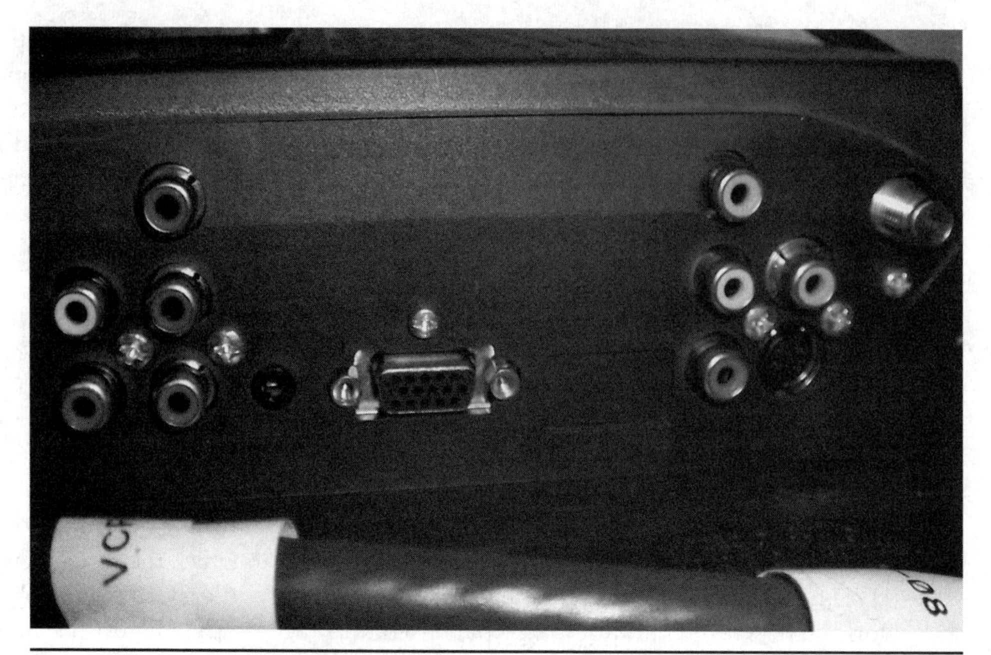

Figure 10-2 Typical TV I/O interface jacks

together in a way that helps you figure out what they are and then connect them correctly with the correct cables with relative ease. The first stack of two RCA jacks are color-coded white and red and are inputs for the audio from the DVD player or other device. The sound standard for component connections is analog stereo sound (not digital), so the white jack carries the sound for left speakers. The red jack is for analog sound for the right speaker.

The component video signals are carried to the screen's internal processor circuitry by the next three stacked RCA jacks and are color-coded, from bottom to top, red, blue, and green in Figure 10-2. When you see five RCA jacks grouped together like this and in these colors, you can pretty much count on that being a connection point for a component connection. The colors of the jacks represent the analog color signals for RGB—red, green, and blue. RGB provides good color rendition and was frequently used early on for high-quality computer-aided drafting (CAD) monitors. I have a DVD player hooked up to the component interface on a TV and the picture quality for DVD movies is excellent.

VGA (Computer) Connectors

The next two connections in Figure 10-2 are side by side—a small $\frac{1}{8}$-inch jack and a larger 15-hole D-subconnector. Here you connect the (analog) stereo sound output and at least VGA quality graphics output from a desktop or laptop computer to the television from the computer's sound card and video cards. The $\frac{1}{8}$-inch jack has three wires and three connections on the little plug. This is a great way to use a laptop or notebook PC at home on a bigger screen. If you invest in a wireless keyboard and mouse, you can sit in your favorite easy-chair and compute from across the room and enjoy the big-screen TV for viewing your computer programs. Wireless keyboard adapters can be found to connect to the PS/2 connectors or to one USB connection on the computer. It is great fun for video gamers in the household to use this option for viewing a computer's output on a big screen. It usually works well if you want any untimely visitors to leave early from your cozy abode—just start showing them family digital photo albums from the computer this way!

Composite Video Connectors

Let's go back to Figure 10-2 to examine the next stack of three RCA connectors. From the bottom to top, they are color-coded red, white, and yellow. These are used for connecting devices and go back to the VCR/early DVD era via *composite* video, so-called because all the color signals for black, white, red, green, blue, and yellow are carried over a single cable. The audio standard for connecting via composite video is analog stereo, so the red and white carry the sound signal for right and left audio channels and the yellow is for the video signals. "Right" starts with *R* and "Red" starts with *R*, so that avoids the confusion on which is right and which connection is not right. All three are usually used, but really old systems have mono sound and will include only one channel connection for audio; in these cases use the white one for mono sound.

Some digital cameras will output pictures to a TV over a composite connection, so it can be versatile, even though it's for an older style connection. If you have an old VCR and a ton of video recordings on VHS tape, you might not want to abandon composite connections, because they will be useful for watching those old tapes. You might also be using a composite interface to connect an old TV with a new digital converter box for capturing OTA digital TV signal programming to your TV instead of over analog channel 3 or 4. The resolution is usually better than that of the analog version.

S-Video Connector

Take another look at Figure 10-2 and you'll see the next two connectors at the right: an orange RCA connector and a DIN style connection for S-Video (which is hard to see in this photograph because it's black). (I'll cover the orange RCA connector in the next section, and the final connection at the far right is discussed in the ANT section.)

S-Video is interesting, because it is one of the connection standards that is a simple single plug-in that carries the video, and you only have to pay attention to the alignment, because the connector is keyed and will fit only one way. Of the four pins or four wires in the S-Video cable, one pair carries the color signal and the other pair carries the luminescence, or color strength, signals. Never force the cable into the slot, because you can easily damage the connection pins. Because the S-Video does not include audio, you must use the nearby white and red inputs for sound.

At this point, you should be thinking "conflict," and you'd be right. On this TV in Figure 10-2, one source component cannot be hooked up using composite and another using S-Video, because of the overlap, or sharing, of right and left sound connections. Of course, if you really needed to make this work, you could use another speaker system to output the sound from one of the devices and use the TV's speakers for the other device.

Coaxial Audio Digital Output

Now back to the orange RCA connector above the S-Video connector in Figure 10-2. On this TV, this is the only "output" connector where the TV is the signal source, and it is for outputting digital sound (actually CD quality digital sound signals) from the TV's tuner to connect to an advanced sound receiver/amplifier that accepts and decodes digital audio. The signal on the orange coaxial connection for audio output is developed from a standard called S/PIDF (Sony Philips Digital Interface).

ANT (Antenna) Connector

The final connection shown in Figure 10-2, on the far right, is the antenna connector that accepts the F-connector end of an RG-6 cable typically from an OTA antenna or cable TV provider. Internally, this connects to the TV tuner that decodes the radio frequency analog or digital signals from the channels you want to watch.

HDMI

High Definition Multimedia Interface (HDMI) is the newest and very popular interface for interconnecting TVs and audio/video devices. Think of the HDMI as the "(almost) everything to everything" interface. Covering the full capacity of HDMI for video and audio device interfacing would nearly take a book of its own. The bottom line for our purposes in our FTA TV project installation is this: If the devices we want to connect together have HDMI interfaces, we should use them if possible, particularly when we want to transfer the signals to see or hear with the highest quality possible. One of the three popular cable connection types for HDMI is shown in Figure 10-3. If you develop an advanced TV viewing experience involving many component A/V source devices, having more than one HDMI input on the TV is valuable to you as the setup person, and from a user perspective, it benefits the entire viewing audience at your residence. Using cables is discussed in more detail in Chapter 5.

FIGURE 10-3 HDMI cable

To help you decipher which jack should connect to which cable, recently manufactured TVs are likely to have a connection key or chart stamped into the case. The connection key is shown in Figure 10-4. The information is difficult to read in the

FIGURE 10-4 Partial view connection "key" molded in TV's case

photo and on the case. When the book for the TV is available, I photocopy the chart, increasing its size a bit, to make it easier to view.

Figure 10-5 shows the only hard-wired control buttons on our TV. As stated earlier, when you're using the interfaces on a TV, you are trying to do a four-way match between the TV's input jack, the correct cable, the source device output jack, and the selection "switch" or menu to present the desired signal to the TV tuner/ decoder. The selection switch can be a physical one or a setting on a menu, or it might be activated with a button on the remote control. If you have ever used a rotator switch to select, for example, a channel on an old TV or selected a speed on a mixer while mixing cookies, you can easily connect the TV decoder or tuner to the right input for viewing the desired source device such as your DVD player with the DVD movie *Despicable Me* loaded into it. On this TV, the selector (rotator) button is labeled "INPUT," as you can see in Figure 10-5. Other labels might be "DISPLAY" or similar labeling. Each time you press the button a different connection is offered up for viewing. When the choices are exhausted, pressing again will bring you back to the top of the rotation. The TV receiver itself is usually referred to as "Tuner" on the display menus.

Regardless of whether you are rotating or pushing, the available inputs are presented as choices one after the other until you return to the top of the rotation and the choices begin again in order. Fortunately, the TV itself gives you some feed-back and shows on the display which choice is current as you progress through the "batting" order. It is up to you to remember or write down which playback device is connected to the various input connections.

1. The antenna/tuner displays an active TV channel.
2. Press the Input (or Display) button the first time from the tuner, and the TV screen will display the word "Composite."

FIGURE 10-5 TV control buttons

3. Press the Input button a second time and the TV screen will display "S-Video."
4. Press the Input button a third time and the screen returns the setting for "Component." As you can see, the DVD connected to component is ready to load a DVD.

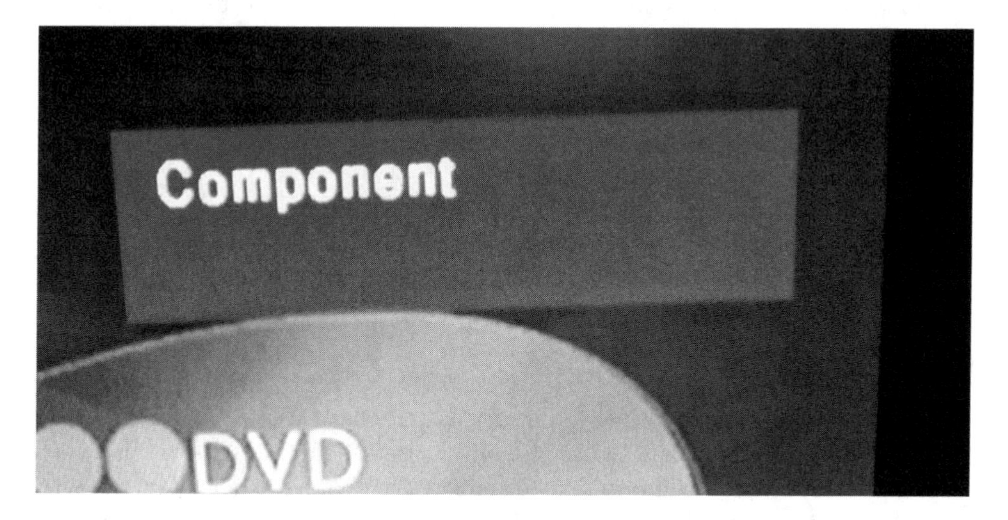

5. Press the Input button one more time and the set shows "PC" (or "VGA") on some TV models.

   If the TV had an input for HDMI, you would see "HDMI" displayed on the screen.

With many of the recently marketed TVs, as you press the Input/Display button or the remote button, the TV highlights the choice directly with an on-screen menu.

When you are done watching the DVD, simply work through the lineup of inputs, making the choices until you get back to where you want to be. Now that was so easy; it's time for another latte.

Web-Enabled or Internet-Enabled TVs

To Web or not to Web is a personal choice. I tend to watch TV on my TV and use my computer to check the lineups on the various networks; otherwise, the TV can be off while the computer is on and vice versa. I like keeping the TV and the computer separate, even in separate rooms. Whether you purchase a WEB TV or set top box for TV screen web viewing, you will have to sort it out, based on your viewing situation and demand for computer and video time and usage.

Portability

Even though we are bombarded with the idea that bigger is always better, even when it's not, some people still appreciate a TV small enough to cart around from room to room or place to place. Bringing the TV to the primary, secondary, or casual viewing area makes some sense if the antenna, input, or cable connections are in place. Maybe you are a weekend warrior camper and take the "garage" TV when you go camping. Decide whether portability is at all important to you. If it is, a compact and lightweight unit will matter at least as much or perhaps more than the other possible features.

NOTE *Digital OTA signals do not work well when you are on the road (moving).*

Ratings by Others

Serious folk out there at institutions such as *Consumer Reports* and many electronics and consumer magazines take a lot of time to test and rate TVs or other consumer electronics. Visit the library to read some of the reviews to get a flavor of what others find important for the ratings. These ratings are, of course, biased by what each reviewer considers important. Your rating scale will probably be different for some features, and something the reviewer rates high (important) might not ever be on your list of essential features. Just remember that when it comes down to making the purchase, you are the only reviewer who counts.

CHAPTER 11

Easy Ways to Integrate DVR, DVD, PC, and VCR

Hooking up video players and recording equipment should be as easy as one-two-three-done. Easy might not be the case, however, if this is your first time. So let us take some time in this chapter to cover some of the basic how-to and what-for needed to simplify this task of connecting audio and visual equipment to and from your television receiver.

The options for connecting video players and recorders are limited in quantity and restricted to certain types of connections based on the available inputs and outputs included on your television and video equipment. Exercise caution in purchasing new audio/visual (A/V) devices to use with your TV and become certain that you have the correct input option necessary for each A/V device. If you are buying a new television receiver and plan to use it with accessory devices, your strategy should be to buy a TV with multiple input/output (I/O) choices. Having more choices is worth paying a little extra for as long as you intend to use the connections. With equipment you already own, check to be sure that you are using the best of the available I/O options.

The basic connectivity on older TVs is limited to a single standard analog signal connection originally intended to receive a signal from an antenna or from a TV cable provider. This is one of the reasons why the government in a form of warped wisdom advised that if you are receiving your TV signal from a cable TV provider, you did not need to buy a digital TV or digital converter box to continue to receive your TV programming from a cable provider. The government advice was true enough but essentially probably qualifies as bad advice for most citizens for two big reasons: it locks the viewer into paying for television programming and denies you the many benefits of using a modern full-featured digital television.

List the TV's Available Inputs

On the very oldest of TVs, the only connection has two screw-on connections intended for a 300-ohm twin down-lead from an outdoor antenna. Many of these TVs are still serviceable and in use, although we might be tempted to think of them as belonging to the dinosaur era. The only option on these old TVs is radio frequency (RF) input. To use the 300-ohm RF input with a piece of modern equipment with a 75-ohm output, say from a VCR or RF modulator, the TV can be equipped with a 75-ohm/300-ohm matching transformer (balun) shown in Figure 11-1 to connect an older TV with shielded cable. The matching transformer will also be used if you're making your own OTA antenna as discussed in Chapter 4.

Most TVs are equipped with a cable-ready connection for connecting RF signals from OTA antennas, cable, digital playback devices, or local modulators. Connecting via analog RF is the least desirable but sometimes the only option on older TVs. You'll find newer equipment that uses digital modulators on the channel frequencies now associated with digital OTA broadcasts. Note that the digital signals from these newer devices and modulators can carry the full definition video and sound of high definition television (HDTV), giving you the best video possible if the TV used is digital and HD with popular screen resolutions of 720p, 1080i, and 1020p. If RF is your only equipment matchup, it is not a bad choice but is potentially an inconvenient one requiring pass-through ports, so you can still receive OTA signals.

Analog TV connections can be made via a digital converter box via pass-through ports if the converter box features them. Figure 11-2 shows an Airlink101 digital converter box's connection ports. This box can connect to the TV via the composite video cord or by generating analog channel 3 or 4 output. When the box is turned off, it will pass through analog RF to the TV's input through its RF OUT connector. Analog modulators with an analog pass-through feature can be connected between the RF OUT and the TV. A digital modulator with a pass-through can be connected between the antenna and the ANT IN on the digital converter box.

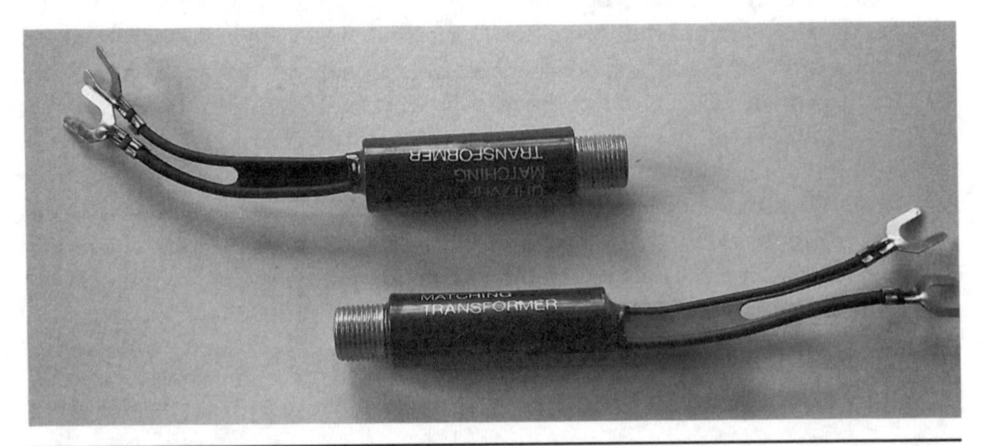

FIGURE 11-1 Matching transformer/balun 300 ohm to 75 ohm

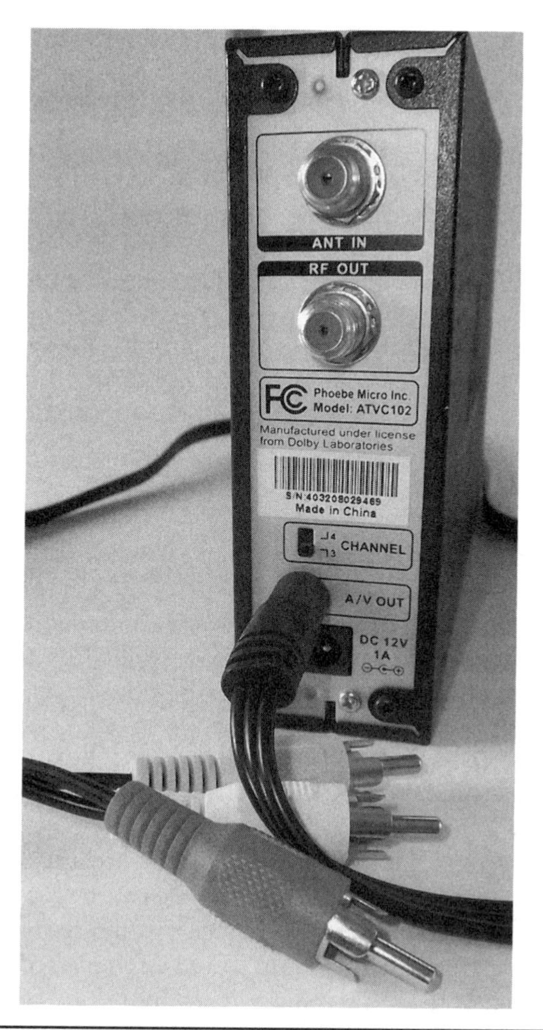

FIGURE 11-2 Connections on digital converter box Airlink101

Determine Available Interface Outputs from the Video Devices

What matters most when you're connecting video devices to a monitor, screen, or TV is extracting the maximum video quality that is possible. To reduce this topic to its simplest concepts, only a few factors influence video quality: The first is the refresh rate, or how many times the picture is refreshed—22, 24, up to 30 frames per second are common refresh rates. Next is luminescence, color, chroma (saturation), and contrast; some video standards are better than others in regard to refresh rates and color quality. Both of these elements are controlled by the original recording, the recording or transmission media, or the equipment that is sourcing the signal, and these factors are largely out of the individual's control once the purchase is made. You do have significant control and influence on the final outcome of both video and sound when it comes to choosing which video connector or audio connection you will use.

The third factor is the number of pixels used to form the image, in both the horizontal and vertical planes. For example, if you have a TV with both HDMI and composite connected to a Blu-ray player that has both HDMI and composite, and you are using the composite connection for Blu-ray, you are not taking advantage of the best possible connection option. The pixel quality rank order possible with the connection type is listed next, with the possible best vertical resolutions on the top of the list.

- **HDMI 1.3** Max resolution 1080p and (2560×1600)
- **HDMI 1.0** HDTV (Full HD) quality, max resolution 1080p and (1920×1200)
- **Computer** XGA 768
- **Component** (somewhat device dependent) 480i, 480p, 576p, 720p, and 1080p
- **S-Video** Standard definition video 480i
- **Computer** VGA 480
- **Composite** 480i 576i
- **RF Analog** 525 lines; technically RF digital TV (DTV) signals (OTA channels) can have any of the above resolutions

Figures 11-3, 11-4, and 11-5 are input and output jacks from the same piece of older equipment—a DVD/VCR combination unit that is typical of thousands of units currently in use.

Connecting this equipment to a television with appropriate ports would have the component output on the player connected to the component on the TV for the DVD's output. The composite on the player would connect to the composite jacks on the TV. The composite VCR input would be connected to an output on a digital cable box output or to the TV composite out for recording programs for later viewing. The stereo audio outputs would be used if there is no digital sound processor to which the DVD audio output can be connected.

At this time, the best quality signals are going to be from HDTV (DVB S-2) satellite channels, then DVB-S channels. For terrestrial TV the quality ranking will be full-channel OTA digital HDTV, followed by subchannel DTV, and then analog TV channel sources. From recorded media Blu-ray will potentially be the best video and audio, followed by DVD, and then VCR tapes. HD-DVD is no longer a viable choice.

Determine the Available Input Jacks on TVs or Intermediate Devices

The challenge here is simply to match the best available jacks on the TV with the best available jacks on the video playback device. Give the video devices with the best video output potential the priority for available ports. Go to the next-best

FIGURE 11-3 RCA I/O connections

FIGURE 11-4 S-Video DVD out connection

FIGURE 11-5 Convenient front RCA connections

video playback device and match that to the set inputs. Repeat the process until everything is hooked up or you run out of ports. When the TV viewing area has a digital satellite receiver, Blu-ray player, DVD, VCR, PC, and a gaming console, the matchup likely would look like this on a TV set with two HDMI inputs, one component, one S-Video, one composite, and one VGA input:

- Blu-ray player connect via HDMI # 1
- FTA receiver connect via HDMI # 2
- DVD player connect via component
- PC connect via VGA
- VCR connect via composite
- Gaming console connect via S-video (sound out to portable speakers)

When two devices compete for best available input, you need to make some compromise choices—for example, a game console that has HDMI might be competing with the Blu-ray player for an available HDMI input. That is where an intermediate device such as a sound processor equipped with video port forwarding might come into play. A carefully purchased sound processor with video forwarding ports can alleviate the jack space shortage by providing a second place to connect audio and video devices. See Chapters 12 and 15 for more on home theater–quality sound systems.

Connect with confidence with what you know and keep in mind that you can always test as you go. Try one device and then the next and compare quality. Make the connections and disconnections while the equipment is turned off and unplugged, and you are likely not to hurt anything if you change your mind later. Getting better video from one device or another is all right—after all, it's your system (but check with your priority viewers first).

CHAPTER 12

Adding Stereo, 5.1, or 7.1 Sound

The objective for adding an advanced sound system is to have the reproduced sound replicate to the nearest extent possible the sound you would hear if you were actually standing in the movie's or program's scene.

Think about how a car horn sounds when you are in a parking lot walking away from your car toward a store entry door. With normal hearing, your senses are first peaked by perceiving the initial attack of the horn's sound and perhaps quickly recognizing the sound as that of a car horn because you have heard car horns in the past. The next instant you would tend to zero-in on where the sound is coming from, perhaps a red convertible a few parking rows behind you and to your left. As you begin to turn your head to look where the sound is coming from, you hear an echo of the sound coming at you as the sound bounces back from the store's window glass and solid brick walls. If other structures are located off in the distance, you might hear secondary echoes as the sound is reflected back to you from many reflective sources. At 70 degrees Fahrenheit, sound travels in air at 1129.5 feet per second. So with solid barriers, such as a cliff's face at 1100 feet from you, it would take two seconds for the echo sound of the horn to get back to you. Because of a phenomenon called the Doppler Effect, the echoes of the original sound from long distances will be altered slightly in frequency (pitch) from the original sound of the car horn. If you were in the scene, your senses would be stimulated by a symphony of reflective sounds propagating in your environment from this one loud event.

There truly is no perfect substitute for being there when it comes to hearing the sounds generated by happenings and events in the real world. Listening in person to a bell choir, hearing the flutes and trumpets of a pipe organ, or being in the audience during a play or musical performance cannot be fully duplicated electronically. Electronic sound reproduction can nevertheless come very close to the real experience with the right equipment—so close that it can become difficult to sense the difference.

Of course, it is totally impossible for a television broadcast or a recorded video to stimulate your senses exactly as they would be if you were actually there in the scene. It is the director's goal, supported by the production staff and sound recording engineers, to bring you as close as possible to feeling as if you were in the program, show, or movie. In the case of music, the idea is to match as close as possible being in the studio, theater, concert hall, or outdoor concert.

Many higher quality sound systems are affordably priced and can enhance your FTA TV and DVD/Blu-ray viewing experiences. When movies or TV shows are made, a lot of production effort is invested in the sounds you will hear as the viewer of a recorded video track. It behooves you to consider matching your home sound reproduction capacity to take full advantage of the sound quality that is included in the digital or analog transmission or embedded in the recording media.

Sound Formats

Reproduced sound included in radio and TV broadcasts or recorded on DVDs or CDs will typically be fixed in the media in one of four common formats: mono, stereo and 2.1 sound, 5.1 sound, and 7.1 sound.

Mono

Monaural sound (sometimes called monophonic) is sound reproduced from a single soundtrack and often from a single speaker. Mono sound is heard from standard AM radio broadcasts, older televisions, and telephones. If it was not so prevalent, it would be obsolete. A mono sound track can be made with the use of multiple microphones so that a wider net is cast to catch the sound, but once blended into one track it rarely does justice to hearing the full quality potential of the sound's original source.

Stereo and 2.1 Sound System

Stereo sound was all the rage when it first came on the scene. Stereo systems record sound onto two separate channels: a left sound channel and a right sound channel. FM radio broadcasts and old eight-track tapes were early uses of stereo sound recordings. Most modern televisions can process and present stereo sound reproductions to the viewer. The left and right channel speakers must be separated by a minimum of 4 feet to offer the best stereo sound. When hooking up stereo speakers, you need to pay attention to the polarity of the speaker and the amplifier connections and connect the positive jack on the amplifier to the positive connection on the speaker. Do this for both speakers. Connecting one correctly and the other speaker incorrectly puts the resultant sound waves out of phase with each other.

Stereo sound quality output is enhanced if a passive crossover is used to channel the bass (lower pitch) sounds to a bass speaker on each channel, resulting in a system with at least two speakers per channel. Additional crossover circuitry can be employed that results in three or four speakers per channel. Stereo sound with a dedicated speaker for bass sounds is called a 2.1 sound system.

5.1 Sound System

In addition to front left and right sound, the five channels of a 5.1 sound system add in the front center sound speaker and a low frequency effects (LFE) speaker, and also left side and right side rear sound channel speakers.

A basic entry-level sound system such as the RCA Model RT2906 (priced at $200 at the time of writing) provides a nice benchmark for adding an inexpensive home theater sound system. Features on this unit include three high-definition multimedia interface (HDMI) inputs and one HDMI output, both optical and coaxial digital audio inputs, and an AM/FM receiver. The unit will allow connection of up to nine audio sources.

AC3 is also a term used to apply to 5.1 sound having six separate audio sound channels.

7.1 Sound System

A 7.1 sound system adds left rear and right rear back speakers to the basic 5.1 sound system and is most common on Blu-ray discs. Sound is carried to the sound processor via an HDMI 1.3 cable on modern equipment.

NOTE *Speaker placement becomes critical for getting the most benefit from advanced sound processing equipment. For speaker placement within the viewing area for 2.1, 5.1, or 7.1, check out this Dolby Sound web site before you plan your speaker setup and placement (www.dolby.com/consumer/setup/speaker-setup-guide/index.html). The web site demonstrates speaker placement for three different viewing distances. Use this site's graphic showing the proper angles as your setup guide for proper speaker placement. Select the viewing distance and this interactive site will return a diagram for proper speaker placement.*

Understanding Speaker Specifications

Although discussions about digital sound reproduction abound, the important thing to keep in mind is that regardless of the media or way the sound is captured or transmitted, all final sound output is still analog, and sound is produced merely by how the sound hits the air from the speakers' vibrations. This reality makes speaker quality selection a very important aspect in the results obtained from any source and sound system. Pay close attention to speaker quality when choosing and setting up a sound system in your home or home theater.

Here are some of the specifications to be aware of when selecting speakers for your sound system.

Impedance

Impedance is a measure of resistance in ohms inherent in the speaker's design, typically four or eight ohms. This is not a choice usually made during speaker selection but is mandated by the amplifier. Each amplifier sound channel expects to be loaded at a given resistance for maximizing the sound output quality. Purchase and install speakers to match the required output resistance required by the amplifier.

Frequency Response

Frequency response refers to the range of frequencies that the speaker can reproduce. A base drum outputs sound in the range of 60 to 1000 Hz and is composed of the initial sound burst followed by decay vibrations. The highest note on a clarinet, for example, is just under 2000 Hz. The harp has one of the widest ranges of frequency outputs with frequencies from 30 to 3000 Hz. A piccolo and violin can reach frequencies in the 10,000 to 16,000 Hz range. The highest "A" note hand bell produces overtones beyond the human ability to hear.

When using just a few speakers in a sound system, choose speakers with the widest frequency range. In multiple speaker systems, crossover (passive or active) circuitry will allow you to select and use specialized speakers intended to produce sounds in a smaller frequency ranges. The whole sound range in a multiple speaker system is then covered by having three or four speakers for each sound channel, each reproducing the sound in its assigned design frequency range. See the upcoming sections "Woofer" and "Tweeter" for more details involving use of multiple speakers.

When purchasing a bundled sound processor amplifier complete with speakers, check the specifications to determine how much of the 20 to 20,000 Hz normal human hearing range is covered by the system you are purchasing. Using speakers capable of reproducing frequencies above 20,000 Hz is thought to improve overall sound quality.

Sensitivity

Speaker sensitivity is a measure of the strength of the sound produced from a standard input of 1 watt of amplifier power applied to the speaker and measured with a sound meter at a distance of 1 meter in front of the center of the speaker. The measurement is in decibels, and the larger the number, the more efficient the speaker is at turning the applied wattage into sound waves. A common measurement would be near 90 dB.

Power Handling

The power handling rating tells you how much electrical input power a particular speaker can handle.

Size

The speaker size typically refers to the diameter of the mounting frame or the widest distance for oval-shaped speakers. Sound quality is affected by speaker size in two ways. Size impacts the volume of air that is set in motion to make the sound. The other impact of size is that it is difficult for a larger speaker cone to vibrate very fast to make the higher frequency sound waves. LFE speakers are large, and small speakers for high-pitched sound are called tweeters.

Maximum SPL

Maximum SPL is the sound pressure level where the speaker will distort the sound.

Types of Speakers

Up to six speaker types can be found in modern sound systems. Single frame speaker systems, often used in radios and TVs, will use either the full range or the coaxial types. Audiophiles will usually have systems that use a combination of woofers, subwoofers, mid-range, and tweeters. Most automobiles are equipped with coaxial speakers when the car has upgraded radios that include DVD and MP3 play features.

Full Range A full range speaker is a single speaker that handles a wide range of frequencies.

Woofer Woofers are speakers intended to produce lower frequency sounds.

Subwoofer Subwoofers are speakers that produce sound in the lowest range of 20 to less than 200 Hz. The LFE on a multichannel sound system is technically a subwoofer.

Midrange Midrange speakers handle the frequencies in the middle range of human hearing capability. In a well-designed speaker system, the midrange speaker will resonate sound beginning where the woofers quit and produce sound up to frequencies where the tweeters begin. Any speaker that resonates sound from 200 to 6000 Hz would be a midrange speaker.

Tweeter The tweeter speaker has a high frequency response and can begin producing sound waves at a few thousand Hz and continue beyond the range of human hearing.

Coaxial Coaxial speakers are often found in car audio systems and contain two or more sound cones of different sizes mounted in the same frame to handle specific frequency ranges.

Crossover *(Note: This is not a speaker type; it is a feature of multiple speaker systems.)* "Crossover" is a term used to describe channeling of sound output to different speakers based on their frequency range. Crossover occurs in two ways: passive and active. Passive crossover is achieved with filtering or blocking after the sound amplifier. Active crossover is achieved with circuitry prior to amplification and requires separate amplifiers for each channel.

NOTE *As a general rule, speakers should not be run at the highest possible power. Applying too much power to the speakers not only distorts the sound output but also generates heat from the voice coil, adding no additional sound volume or quality. When purchasing speakers separate from the sound processor/amplifier, buy speakers that will handle slightly more power in wattage than the amplifier will produce per channel.*

Connecting the Audio Outputs

Selecting the best audio outputs from the video playback devices to connect to the sound processor/amplifier is fairly straightforward. The quality hierarchy in general is HDMI 1.3, HDMI 1.1, followed by optical, coaxial, stereo, and mono if it is the only available output. In the next five illustrations of the business end of some example audio equipment, we will select the best audio source to connect to the audio processor's available audio inputs. The equipment you have might not be exactly laid out or labeled the same way as these samples, but the selections from your unique playback equipment will likely present similar choices.

In this first example of video playback equipment, you can see three choices for connecting the audio to a sound processor. The best selection would be to use the HDMI Out to the sound processor as long as the sound processor can forward the video signal to the TV set. The next best choice would be to use the Coaxial Digital connection. The last choice would be the L and R analog audio. This last choice would yield only stereo sound from the processor.

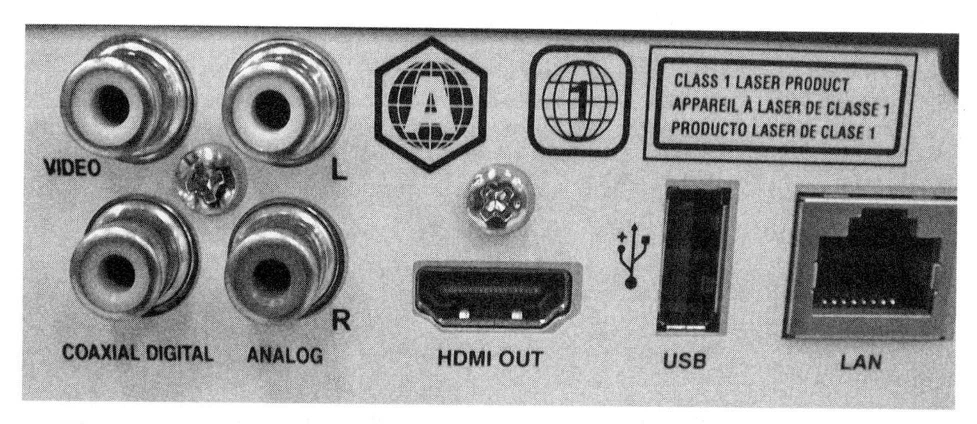

The next example includes an option for both Component and Composite Video output (the composite output is simply labeled VIDEO and is color coded yellow) and has essentially the same options for Audio output; HDMI, followed by Digital, followed by L and R stereo outputs, in order of preference for use.

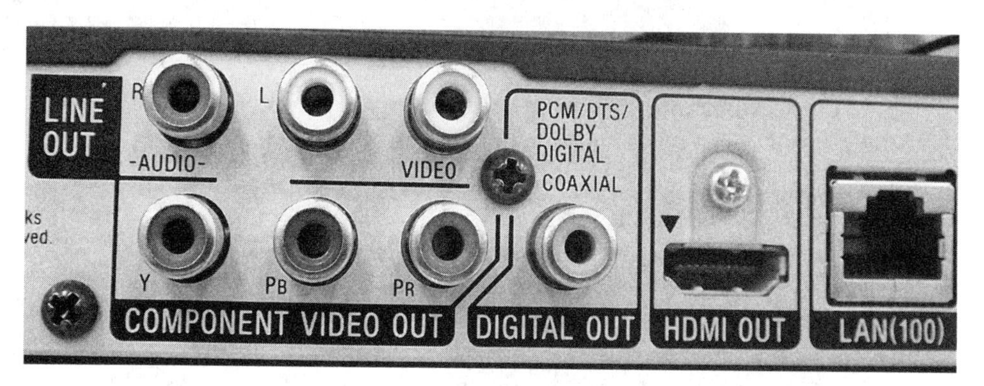

The next playback device includes output from VCR tapes and DVDs. For sound output to the sound processor from the DVD/VCR section, the only choice for use is the L and R stereo jacks. The DVD player side offers a digital audio Coaxial connect that would normally be the best choice for cabling the sound output from the DVD to the processor. The L and R stereo jacks would be used if this is the only option for connection to the sound processor.

The next illustration offers in order of preferred use: HDMI Out, Coaxial, and the L and R stereo for connection to the sound processor.

In the final example there is no HDMI connection, so the preferred connection in this case is the Coaxial Digital Out jack if the sound processor has this input available.

In a typical home system, the devices usually connected to the sound processor amplifier are a DVD/VCR combo unit, a stand-alone Blu-ray player, and the FTA or paid-for service satellite receiver. Always use the best connections available from the payback device to connect to the processor. Using the L and R connections from a Blu-ray player, for example, is not providing the processor with the best sound possible. If you have to compromise for lack of available quality connections on the sound processor, use the best first for the FTA receiver, then the Blu-ray, followed by DVD, and VCR last. Stated another way, the FTA receiver, Blu-ray player, and DVD should be connected via HDMI or Coaxial connections whenever possible.

Adding quality advanced sound simply involves paying attention to the details for the sound system's specifications and using the best available connections to an available input on the sound processor/amplifier. Connecting these few dots together will take you from so-so sound to sound that brings you and your home viewers as close to being into the scene itself as possible. Big screen TVs, for example, are best enjoyed when high quality sound is also an essential part of the viewing experience.

CHAPTER 13

Installing a Satellite Card FTA Receiver/ Tuner in a PC

Microsoft and computer manufacturers have teamed up to promote the potential of the personal computer as the centerpiece of the home theater or entertainment center. With a computer or two equipped with FTA cards and Network Attached Storage (NAS), plus broadband/DSL Internet access, there is virtually no limit on the amount of songs, movies, and video clips that can be stored and later accessed and played. With NAS, more than one computer can access the files to play the movies or stored programs, making this a very desirable method to provide viewing and listening options throughout the home.

For computer aficionados, using the computer can be second nature, and the idea of having anything and everything just a mouse click away is...well, normal. Software, add-in cards, peripherals, and PC system boards are constantly becoming more suitable for making the computer the center for entertainment and education at home or in the office. At the same time, other audio video devices and TVs are becoming more "computer-like" and sometimes easier for the novice to operate than a computer might be.

Using a computer for viewing FTA programming as a one-off or making it an integral part of a home entertainment center is not a big leap when you consider that the dedicated FTA receivers are frequently using open source computer operating systems embedded in the firmware of the receiver. Whether you choose to extend FTA to the computer or extend the computer to be part and parcel of the home entertainment center, reception of FTA programming can become an everyday experience without emptying your pocket.

This chapter discusses the addition of a satellite TV card to a desktop or tower TV for those who like the idea of expanding the value and usefulness of their computer by making it the control center for capturing FTA satellite TV channels and programs. First some details covering the possibilities and products that are needed.

Know the Candidate Computer Intended for Adding an FTA Card

Personal computers have long used standard interfaces, which has allowed multiple manufacturers to produce components, cards, and peripheral equipment to work with the PC and the various generations of operating systems. The "heart and lungs" of the computer are the processor and the system board.

System boards have expansion slots for adding cards such as network interface cards (NICs) and advanced video cards that extend the ability of the computer to perform additional functions. Over time, popular cards are often integrated into the system board's electronics. Before you order and add any type of TV card to a computer, you need to know whether the system board has available expansion slots for adding a card and what type or types of expansion slots are available. Early computers included only an ISA (Industry Standard Architecture) 8-bit expansion bus card, and then the EISA (Extended Industry Standard Architecture) was developed to extend the number of add-ons possible. Later on, the Peripheral Component Interface (PCI) was developed, and then came the PCI-e (express) in several versions, with the newer versions of the standards often being "backward compatible," meaning that old cards would work in the new bus card slot.

Know the Candidate FTA/TV Card

TV cards are currently available for PCI and PCI-e, and small plug-in models are available for the USB port intended for use on portable computers. Once you know what expansion slot type the computer has available, you can then shop for the TV card of your choice.

Two types of TV cards are produced: those intended for OTA (also referred to as terrestrial TV in Europe) and those intended to receive and decode digital FTA satellite broadcasts. For the two terrestrial digital tuner cards, two types of standards apply to television broadcasts, one for the United States and the other for Europe.

The following acronyms are used to differentiate the cards in the ads and sales literature:

- **DVB-S** Standard definition satellite
- **DVB-S2** High definition satellite
- **DVB-T / DVB-T2** Terrestrial VHF/UHF OTA TV tuners (The DVB-T designation is not necessarily enough to get you the card for OTA in your country. DVB-T is a standard in Europe for OTA antenna broadcast station reception, and digital ATSC is the standard in the United States and associated countries.)

Broadcast Standards Supported

To find the right card for your region and purposes, you need to dig a little deeper in the card specifications. What you look for in supported standards depends on where you live and what you want to receive and decode. Here is what to look for:

- **NTSC** National Television System Committee standards support decoding of analog cable TV and analog OTA TV.
- **ATSC** Advanced Television Systems Committee standards support digital OTA TV in North and Latin America.
- **QAM** Quadrature Amplitude Modulation standard for carrying digital cable TV channels; a QAM tuner in a TV or card allows reception of unscrambled cable channels without a set top cable box.
- **DVB-S** Digital Video Broadcasting–Satellite standard supports standard definition digital FTA signals worldwide.
- **DVB-S2** Digital Video Broadcasting–Second Generation standard supports high definition digital FTA signals worldwide.
- **DVB-T** Digital Video Broadcasting–Terrestrial is the standard for terrestrial TV for a European-based consortium that includes Europe, Australasia, and parts of Africa and Central and South Americas; DVB-T2 is the second generation.
- **ISDB-T** Integrated Services Digital Broadcasting–Terrestrial standard supports digital TV broadcasts in Japan.
- **DMB-T/H** Digital Multimedia Broadcast Terrestrial/Handheld standards are used in China.
- **DAB** Digital Audio Broadcasting is a worldwide radio digital audio standard.

NOTE *Some manufacturers have developed PCI cards with multiple input jacks for cable connections: one for connecting to two or more sources such as cable TV, one to connect to a FTA low noise block (LNB), and another for an OTA antenna or to an FM radio antenna. In the United States, a multi-input card that supports FM, NTSC, ATSC, QAM, and DVB-S (and DVB-S2) would be ideal, if it existed.*

System Requirements

Each card lists certain system requirements for the computer hardware and also addresses matching software operating system requirements and codex levels. The computer must at least match or potentially exceed the card's listed system requirements. Exceeding the OS software revision level can be problematic, because newer versions of the OS will not always be compatible with the older hardware. As products move out of the supply pipeline, the hardware-linking drivers are often unavailable for the cards or devices to work with the newer computer operating systems. Following are some example system requirements:

- **Processor and/or minimum processor speed** Intel Core 2 Duo 2.8 GHz
- **Operating system** Windows 7, Mac OS X
- **Available slot type** PCI or PCI-E
- **Graphics card or minimum card memory** 128MB
- **Sound card** Sound Blaster
- **Drive support** CD-ROM drive
- **Codex** Directx9.0c

Card Features

The requirements for your selection of an FTA card will be nearly the same as the features you would seek for determining which FTA receiver to purchase. Chapter 6 offers details regarding receiver features, and Chapter 10 covers many of the highlights you can use to select a TV set—some of which have relevance for selecting an OTA-capable tuner card. Some neat features available on some of these cards deserve mentioning here. All the cards should push received video to the computer screen, and some offer outputs for sending the video to a TV screen or other device; some additional functionality adds value to the viewing experience.

- One nice feature is putting video output to a separate window on the computer while you do your homework in another window. A picture in picture allows listening and seeing the news on FTA, and seeing the ball game in a smaller window can be handy for the multitask-oriented individual.
- You can make the computer work like a DVR, catching programs while you're away from home by recording digital satellite TV to the computer's hard disk or network drive in MPEG-2 format for later viewing. MPEG TV recordings will use about 1.5GB of disk space for each hour of recoding. Pause and replay are also useful features.
- Support for the DiSEqC protocols for switching LNB inputs could also be important.
- Look for a card that supports USALS for controlling dish motors.
- Supplying LNB 13/18 voltages is necessary in the specs for FTA LNB polarity switching between horizontal and vertical.
- When the computer is included in your home theater setup, having an IR (infrared) remote control feature is almost a necessary requirement. Computers equipped with the latest home theater software will often have their own remote control. Using two remotes might be necessary with some TV cards.

Installing a PCI-e Card in a Computer

One of the two prototype cards discussed in this chapter is the LifeView FLY TV-EXP LR860STAFR-R PCI-e card. The OTA antenna connection on this card is intended for European terrestrial broadcast, so using this card in the United States for OTA would require using a digital converter such as the Airlink101 hooked to the analog input or simply accepting that it will do FTA and analog, leaving OTA out of the equation. At under $50, the card still presents a value proposition, even if it's used only for FTA. You can also install a card that is dedicated to function only as an OTA (ATSC) tuner in the same computer alongside a dedicated FTA card. This particular card is not supported on the Windows 7 operating system.

Figure 13-1 shows all the items that were included with the card, which means you don't have to hunt down some of these specialized items. The most important of the kit's components would be the installation CD and software drivers for the card.

Among the unique features on this card are the four input jacks for cable inputs to the various tuner features built into the card. Looking at the connectors on the left side of the card shown in Figure 13-2, from the top, is a connector for the down-lead from the satellite dish's LNB. The next connector is for reception of European DVB-T via an OTA antenna. The third connector is for the FM radio tuner antenna, and the bottom connector is for analog television or analog cable connection.

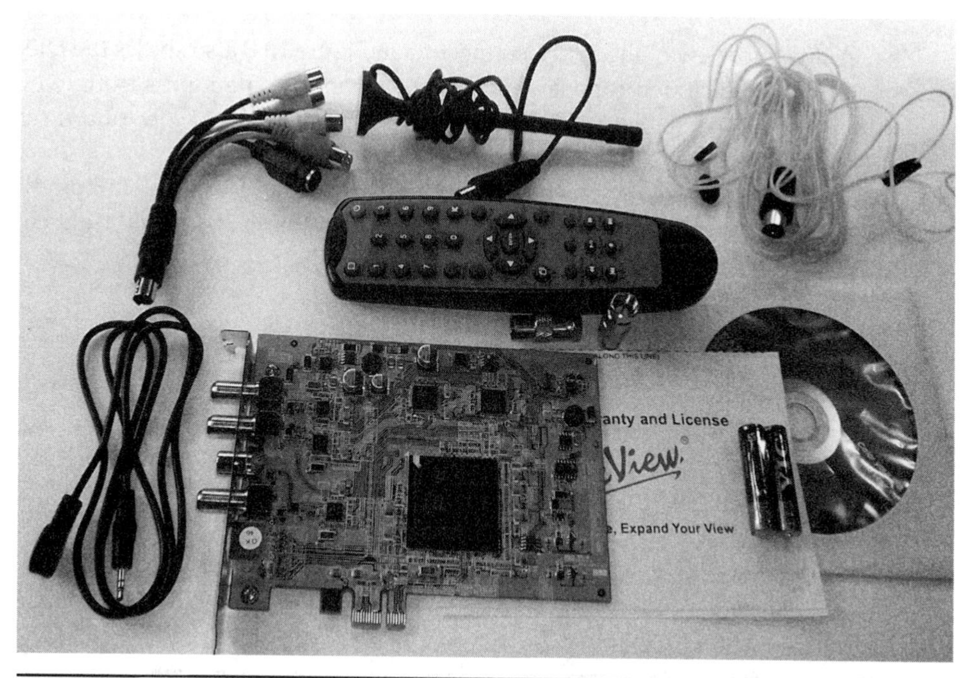

FIGURE 13-1 PCI-e satellite tuner card and accessories by LifeView

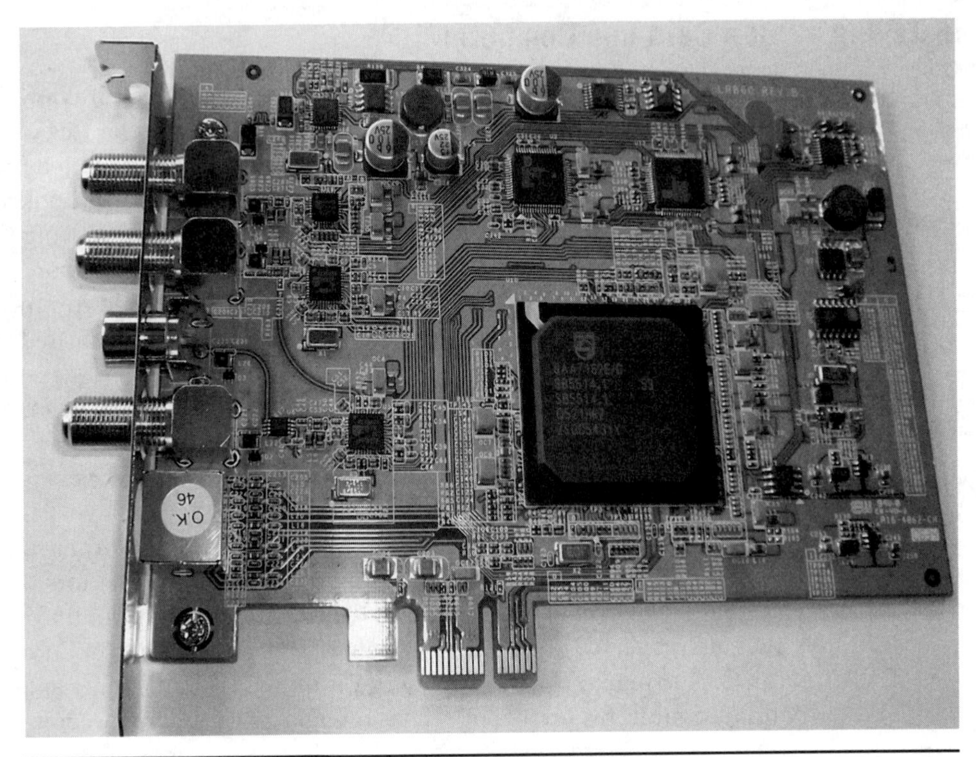

Figure 13-2 Side view of FTA card

A close-up view of the card's connectors in Figure 13-3 also shows the DIN connector for the included cable assembly, which can provide inputs from two other audio video sources such as a DVD player via composite or S-Video outputs on the players.

The included cable connection assembly is shown in Figure 13-4. From the left are S-Video, composite video, remote infrared, and two each left and right channel stereo sound connectors.

Figure 13-3 Connector end of FTA card

FIGURE 13-4 Custom cable connecting jacks

A close-up of the card connector in Figure 13-5 shows the two small tabs included in all standard PCI-e cards. The two tabs are inserted into the PC system board, providing the communication between the card and the system memory and main processor.

FIGURE 13-5 Close-up of PCI-e system board connection tabs

Also included in the kit are two F-connector to slip barrel adapters for connecting the FM antenna or the small digital OTA antenna included in the packet. If external/rooftop antennas are used, the connections would be made with RG-6 cable with F-connector ends connected directly to the jacks on the card. The FM antenna's barrel connector is shown in Figure 13-6 between the two adapters.

Figure 13-7 shows the infrared receptor bulb and cable for interfacing to the remote control included with the kit. Use double-sided sticky tape to hold the sensor bulb toward the open room where the remote will be used.

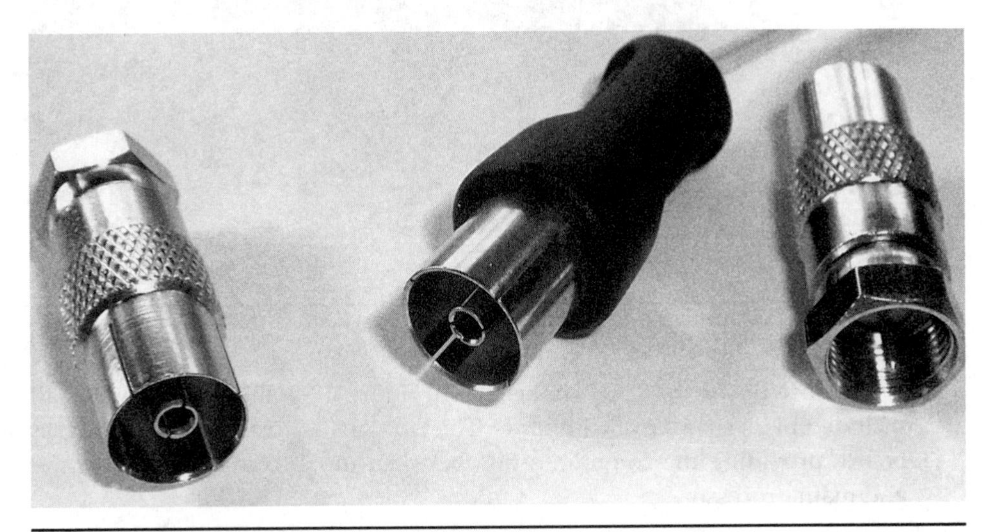

FIGURE 13-6 F-connector adapters

FIGURE 13-7 IR remote cable

To test a remote control to see if it is emitting infrared light, simply hold the remote up to a digital camera and press the remote's On button. The digital camera or cell phone camera will pick up and display on the camera's LDC the infrared light that cannot be seen by the human eye. The next illustration shows a working TV remote before the button is pressed.

The next illustration shows what the camera will show when the On button is pressed. If the remote does not work even after the batteries are replaced, try this.

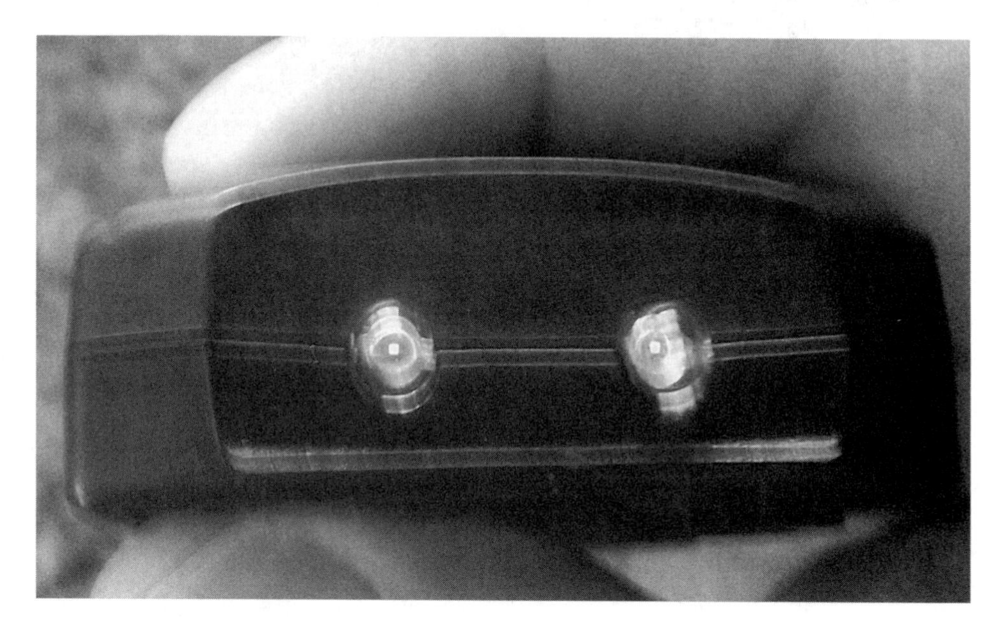

TIP *If you ever need a new remote for an older TV or player, call 1-800-Remotes. They might be able to help you find one.*

Before you install a PCI-e card in the computer, start by placing the computer on a bench or tabletop. Disconnect the computer from all electrical power, and disconnect monitors and peripherals from the computer. (Think no strings attached: no wires.) To decrease the possibility of your encountering static electricity, avoid doing this work on a carpeted floor and use a grounding strap wire from your wrist to the computer case. Radio Shack stores frequently stock "static control wrist straps." Using one can prevent damage to sensitive electronic components from the ravages of static electricity.

1. Open the computer case to reveal the system board. If it is not obvious how to open the computer's case, look in the service manual for the computer or go online to find one. Manufacturers use differing methods to fasten the computer case covers, including slides, screws, or snaps.
2. After the case is open with all the cords removed, connect the grounding strap to the case and your wrist. Then it is OK to remove the PCI-e card from the bag.
3. For a Compaq tower computer, you loosen one screw to remove a slide cover exposing the system board. To "open" a slot, you need to remove the clamp on the outside of the computer that holds the slot covers in place, as shown in the next illustration.

4. The two PCI-e expansion slots into which the card's tabs slide are shown in the next illustration. The PCI-e slots are rectangular and raised above the system board. Notice the separator on the socket between the short and long tab socket. With the slot cover removed, remove the metal filler for that slot and simply slide the card into the system board so that the card's tabs are fully seated into the PCI slot and the card's back bracket is in place ready to be clamped.

5. Replace the clamp on the back of the computer and put the remaining slot covers in place. On some computers, individual screws hold each slot cover in place; on these models, use the screw to hold the card in place. Check the computer's manual again for particular instructions on how to hold the card in place.
6. Replace the cover and reconnect all of the cables, hooking up the power cord last.

The installation from this point on will vary according to the instructions provided with the specific PCI card you are installing. For most, your next steps will be to boot up the computer and load the installation CD, following the instructions that appear on the screen. When following the install prompts, you'll probably be asked to answer the installation questions in the affirmative and accept the defaults of the installation program steps.

Once the card, software drivers, and software programs associated with the card are installed on the computer, power down the computer, unplug it, and connect up the cables to the card inputs. Take care to connect the LNB cable down-lead from the dish antenna first. Information on making down-lead cables, using switches, grounding, and surge protection is provided in various other chapters in this book. Setup and stocking of channel view lists on the PC will be similar to doing blind scans on a FTA receiver box as well.

Hauppauge makes a dedicated FTA tuner card—Models WinTV-NOVA-HD-S2 and WinTV-Nova-S-HD—that fits into a standard PCI slot and is supported on Windows 7, Vista, and XP with SP 2. Packaged with the Hauppauge card are the items shown in Figure 13-8. The WinTV-NOVA-HD-S2 receives and decodes both DVB-S and S-2 signals in SD and HD programming from satellites, and the WinTV-NOVA-S-HD receives and decodes only DVB-S broadcast signals. Both models support DiSEqC version 1.0 for switching to multiple LNBs.

FIGURE 13-8 The Hauppauge WinTV-NOVA-S-HD package

Notice the difference on the PCI bus shown in Figure 13-9 from the PCI-e bus connection tabs pictured in Figure 13-3. If you have a recently purchased Windows-based desktop, tower, or mini-tower computer, it will probably have one of these two bus style slots available.

Figure 13-10 shows the connector (back) end of the card, which has inputs for the satellite down-lead, an S-Video input, composite video input, one stereo line-in for audio, and a small jack for the IR (remote) input. These inputs allow you to

FIGURE 13-9 Close-up of PCI bus on Hauppauge satellite TV card

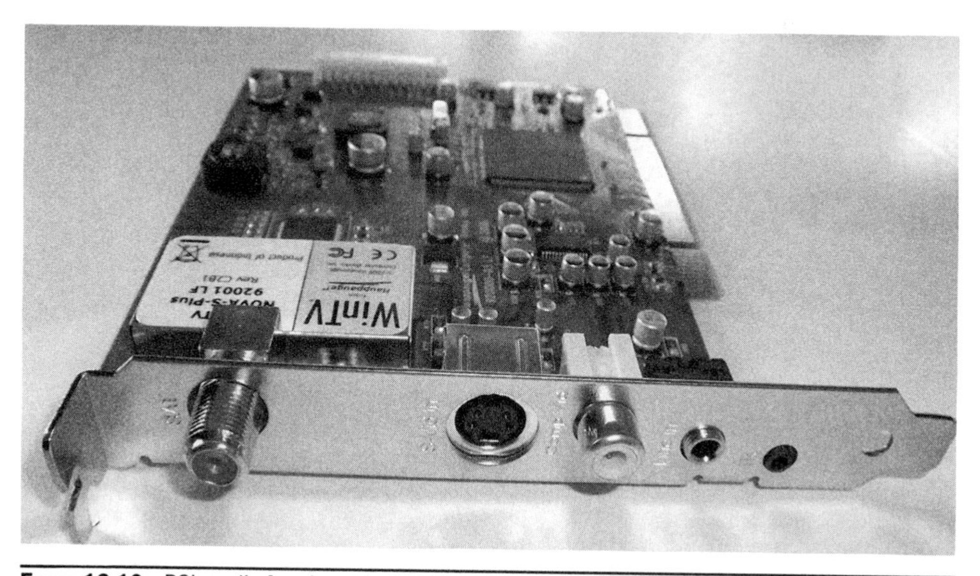

FIGURE 13-10 PCI card's four input jacks

display video from other devices. The software included in the Hauppauge disk will allow for later playback of programming saved to disk.

With a UK price of £83.28 for the Hauppauge PCI card, it should be possible to buy one in the range of U.S. $150–200. The installer program WinTV v7 for the Hauppauge card is written so that you can choose any of 20 languages for the installation and user interface. The program will take you through installing the drivers and then installing the WinTV setup program. Hauppauge also manufactures an array of other interesting video and television cards for video recording and viewing.

CHAPTER 14

Watching TV, Movies on Demand, and Video over the Internet

Remember when going to the video rental store was all the rage? Video rental stores sprung up all over the country like dandelions on a green lawn. You could find movies recently released on VHS/DVD for a pittance of a fee compared to what you'd pay at the theater. And often at the corner grocery, along with the latest released video, you could pick up soda pop and popcorn, and my all-time-favorite movie munchies, chocolate-covered raisins. We can still rent DVDs at small shops today, and even at many public libraries, but with the availability of the Internet, a monumental shift has taken place.

It is great to have choices for entertainment and educational video programs. We want viable, workable, and inexpensive alternatives to get the programming we want to watch, when we want to watch it, and in the comfort and convenience of our own cozy little cottages. The fewer hassles the better, and as great as the video stores are, they do not always have what we want to see, and petrol is not getting any less expensive. This chapter is about adding to those choices after your main FTA TV viewing schema is in place and fully operational.

Most fee-based satellite programming providers charge at least a minimum monthly fee for service that includes a base line of channels and, depending on where you live, local channels. After the initial sweetheart deal for a few months, you'll pay more to access these channels and the provider's library of movies. The company will want you also to have a broadband (Internet) connection to connect to the receiver. This means added costs, unless you are fortunate enough to live in a neighborhood with free WI-FI Internet service. If you have no broadband connection, you will need to connect a land-line phone to the receiver. Many consumers have given up on land-line phone service these days in favor of more useful cellular phones and the associated services cell providers offer.

NOTE *If you were to call the major fee-based satellite programming providers in the United States and ask what it would cost to get movies on demand service from them, you would find that, among other things, it is nearly impossible to get right through to a pre-sales customer representative by phone. You would also find out that the automated prompts for the voice response system for both companies qualify as "brain dead" in this author's opinion, unless you are already a customer. Give them a call just for fun and find out for yourself.*

You readers are the smart guys and gals because you have installed (or are planning to install) one or more FTA receivers with one or more satellite dishes attached, so you can get a wide variety of English language and international programming for—drum roll please—*free*. You have added an improved antenna system for OTA signals to maximize the number of "local channels" you can also receive—for *free*. Now perhaps you would like to watch a movie from time to time without leaving the house or breaking into the savings jar when the local library is closed. To do so, however, will also involve having a decent Internet connection and the expense associated with getting the service.

After you have secured an Internet connection at decent broadband speeds, you have a number of options for viewing video and listening to audio sent to your TV set or computer.

NOTE *None of these viewing options or entertainment sources is endorsed by the author or McGraw-Hill. They are included here only to present examples of what is possible and available for expanding your at-home viewing options. Simply stated, make your own choices for alternative viewing sources. This chapter offers just some of the possibilities.*

Getting Internet Service to Your Residence

Depending on what study you read, about 80 percent of U.S. homes have Internet access today. Regrettably, as many as 25 percent of those homes are on POTS dial-up connections that limit users to narrow bandwidths between 14.4 and 56 Kbps, and that's not nearly enough bandwidth to download video at a substantial enough rate for smooth viewing. If you do not have an Internet connection or are still using dial-up, you will have to consider the incremental cost of subscribing to one of the faster broadband Internet services if you want to watch streaming movies.

Because about 96 percent of U.S. homes have or have had telephone service over land-lines connected at one time or another, the best option for going to broadband speed Internet service might be to use digital subscriber line (DSL), or Asymmetric DSL if it is available. If you live too far from the main telephone switching station, DSL might not be available. Many companies, including AT&T, offer DSL even if the household has abandoned land-lines for phone service in favor of cell phone service as so many have done. DSL is installed by using a phone jack filter that plugs into the telephone wall outlet and provides one noise-filtered jack for the phone plug and the other unfiltered for the DSL modem up-link connection. The registration process is usually more time-consuming than the actual hook-up. Using DSL is easy, because most homes and apartments already have phone outlets in one or more rooms, so additional wiring is probably unnecessary.

Another easy option is to choose a cell phone provider that allows broadband speed connections to the Internet from your home location over your computer. I travel in my motor home most of the year and find that my Palm-Pre cell phone (handheld computer) from Verizon provides a WI-FI hotspot for up to five computers to connect and offers download speeds in most locations sufficient enough to enjoy video viewing on the two computers we take when we travel. I use it every day, everywhere we go. Although I would not give it perfect grades for connectivity performance, it is sufficiently fast and reliable in nearly every locale.

Many different phones are available from all the major cell carriers, and a small percentage of them will allow corded (tethered to the computer) or dedicated wireless device connection to the Internet over the cell towers. These other device options include cards, USB cards, or hotspot packs that work independently of a phone. If you go this route, check to see if the plan includes unlimited downloads or what the limits are before you sign up. The Internet access portion of my cell service currently costs about $30 plus tax each month. With movie theater tickets currently at $6 to $10, if I used the cell phone to download only three or four movies each month, it's a wash on the cost. In reality, I use the cell phone's Internet access for e-mail, Facebook access, Google maps, data transfer, Internet searching, and much more. Plus it is fully portable to go on the road. So for me it presents a great value because of all the things it will do in addition to providing the Internet connection for my two computers. Dick Tracy's 1940s era wrist radio was certainly prophetic of the technology to come in portable communications. With 3G and 4G cell service, it only gets better. The only fear is that it might get more expensive to keep up on speeds to stay connected.

Speedy Internet Connection

Connection to the Internet that sustains download speeds in excess of 384 Kbps is the absolute minimum for watching small video clips. When streaming video over the Internet, your computer or devices will buffer the bits being downloaded if your connection speed is not fast enough. Buffering means that the bits are stored in the memory chips or on the device's hard drive until there is enough data to play to the screen and sound card. If you are trying to watch a lengthy video and your Internet connection is not fast enough, the picture will freeze while the buffering process is repeated or you will see flutter (constant stopping and starting) of the video.

Having just the bare minimum speed is not the desirable way to go. Video clips can be recorded and played back in a number of different screen resolutions or formats expressed as the number of screen pixels in the horizontal by the number of pixels in the vertical. There are also different encoding formats, with MPEG-4 being the current and most popular for quality. For the formats listed here, you will need a connection that is able to maintain bandwidth through-puts near those listed:

- **Standard video** Resolution of 320×240 pixels; 500 Kbps or better bandwidth
- **High quality** Resolution of 480×360 pixels; 1 Mbps or faster download speeds
- **HD video** Resolution of 1280×720 pixels; 2 Mbps or faster download speeds

Shopping for Broadband

When you shop for an Internet connection from broadband providers, include your local phone service provider and/or the local cable TV company in your search to see what is available in your neighborhood. Unfortunately, most of the cable companies want to sell you the whole bundle—cable TV, IP phone service, and Internet, which increases your monthly fees. This makes offerings from major telephone companies such as AT&T such a great value, where you can get ADSL Internet-only service on a phone line (without any phone service at all) for less.

If you already have a speedy and reliable Internet connection, you can use third-party services such as Skype (www.skype.com) or magicJack (www.magicjack.com) to make phone calls if your upload speed is at least 128 Kbps. For those in rural areas, choices might be further limited to using a cellular phone service for broadband or a satellite Internet provider. Ask the cellular companies that service your area if you can try a phone at your residence before you commit to a contract.

If your residence is a long way from the cell tower, some phones can be tethered to a computer to provide broadband and can also be connected to an external (outdoor) antenna to improve service. Wilson Electronics (www.wilsonelectronics.com) offers more information on antennas and cellular signal boosters. The upload and download speeds will be different and the available packages will be tiered with the highest bandwidth number being presented for the optimal connection. None of the providers will be able to maintain the optimal numbers for bandwidth throughput all of the time.

With DSL, you will find plans with download speeds typically incremented at 768 Kbps, 1.5 Mbps, 3.0 Mbps, 6.0 Mbps, 7.0 Mbps, and 15.0 Mbps. With fiber-optic services to your home or business, the speeds can range from 15 to 50 Mbps. With satellite Internet service, the download speeds will be incremented also at rates similar to these: (up/down) 512 Kbps/128 Kbps, 1 Mbps/200 Kbps, 1.5 Mbps/256 Kbps, and 2.0 Mbps/300 Kbps.

Internet Protocol Communication Device

Once you have a decent connection to the Internet, you can enjoy the benefits of being able to download videos, movies, music, and educational information from the Internet and watch it on your TV. You'll need some sort of Internet-enabled device that can make a connection out to your TV. The device you use must be able to do two things: It must communicate over the protocol (TCP/IP) used on the Internet to download video files, and it must process those files with video and soundtracks and output them to your viewing and listening device, aka TV. So it has to output sound and video over a component, composite, VGA, or HDMI type connection, and match at least one similar connection to your TV.

Your computer or laptop, for example, would connect via a VGA cord and a ⅛-inch audio jack. An Internet-enabled Blu-ray Disc player will connect best over an HDMI cable. Check your connections to make sure you have a match. The device can connect directly to the Internet router via a CAT 6 patch cable (covered in Chapter 5) or over WI-FI. If the device does not support wireless natively, you might be able to use a wireless bridge. (It seems that with every International CES consumer electronics tradeshow, sponsored by the Consumer Electronics Association, another new device that communicates over IP is available.)

Desktop Computer

When it comes to processing video and audio over a desktop computer, the beefier the system the better. Think in these terms:

- Windows 7 Home Premium (64 bit) as your minimum OS or the latest Apple OS version for Mac computers
- Processor speeds near 3 GHz
- 3GB of installed memory
- A large hard drive of at least 400GB
- A video card and monitor that supports 1280×720 or 1920×1080 resolutions and is backward compatible with VGA and has a VGA connector.

Laptop Computer

When it comes to processing video and audio on a laptop computer, again the more power and performance, the better the device. Think in these terms:

- Windows 7 Home Premium (64 bit) or the latest Apple OS
- Processor speeds of at least 2 GHz
- 2GB installed memory
- Audio codex that supports six channels

- A large hard drive of at least 400GB
- VGA connector; some small notebook computers may have no video outputs

You can up the cost by finding computers or building one for yourself with extra video and audio bells and whistles. One neat feature for a desktop or tower PC is a sound card with 7.1 sound outputs built right on the PC, such as the one shown in Figure 14-1. From the top left is a jack for digital audio in, a stereo sound out (front speakers), and a stereo microphone in; from the bottom left is an output for the center/subwoofer, for rear speakers, and for side speakers.

You might have a computer with many of these features built in. It is common for computers to have built-in features that users do not always use. Three things are worth the extra cost to give you a better viewing experience.

- Great video cards that can handle fast refresh rates, and of course a matching monitor or TV that supports those rates is number one.
- An audio sound processing card that supports 5.1 or 7.1 sound system processing is always a nice feature. A six-channel codex supports 5.1 digital sound processing, but you must have audio equipment and the required number of speakers that support it to get the full benefit of the sound system.

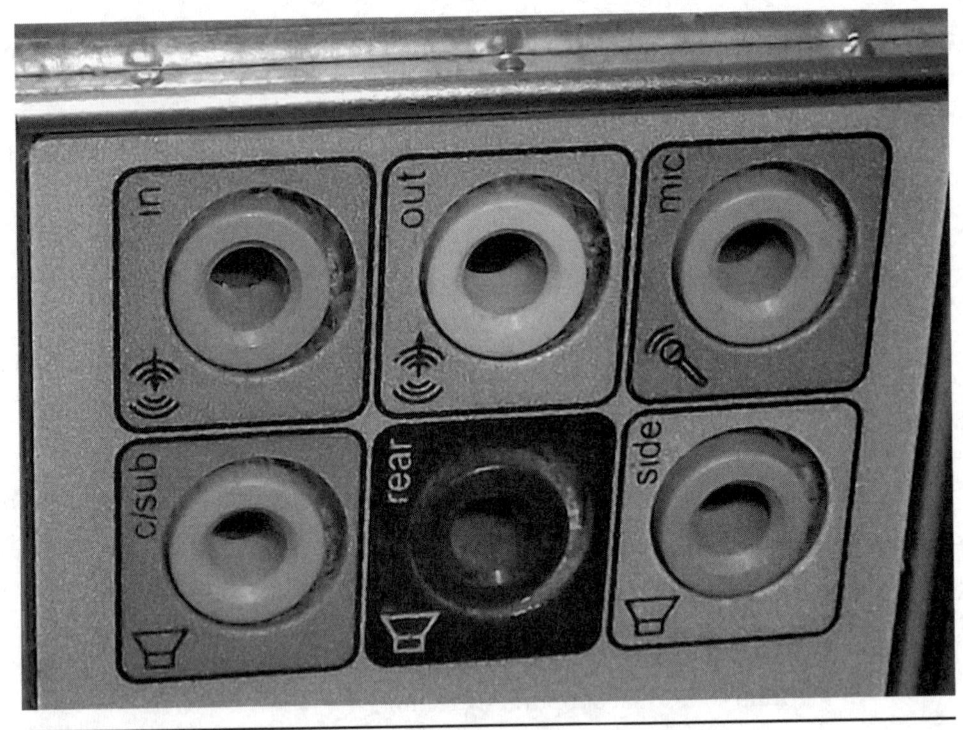

FIGURE 14-1 PC sound card connection jacks

- More storage (hard drive space): the more you have, the better for storing videos and movies for later viewing.

NOTE *You can also archive your audio library on hard drives. It helps to have a backup plan and back up your files (video/audio/photo collections) periodically when your library is large enough to warrant the effort. There are easy and inexpensive ways to do this, such as networking an old computer, buying network storage, or using portable USB drives for backup. For a home or small office computer system, using extra USB drives simplifies the backup process.*

Blu-ray Disc Player (Internet Enabled)

Blu-ray offers fantastic graphics and video output. Watch an action thriller on a Blu-ray Disc and you can easily get hooked. Video over the Internet will not give you the same quality of video and audio performance as a Blue-ray Disc, however, even if you are using a Blu-ray Disc player. In addition, players might not work with or be fully compatible with every available on-line video source, which runs on web servers using many different operating systems and file types. Look at the latest offerings for disc players, and compare features and price before you make your purchase. Take a look at these manufacturers at a minimum to find the best value: Sony, Panasonic, Samsung, and LG. Compare these top-name brands for price and features to other manufacturers of Blu-ray products you might be considering. You can also add a Blu-ray Disc reader or reader/burner to a computer for playing or even making your own Blu-ray Discs. A disc reader/writer that includes LightScribe technology will also "burn" professional looking labels on the flip side of the disc.

Wii

Wii gaming seems to be popular with all age groups. A Nintendo Wii can connect to the Internet via a broadband connection and download content from on-line sources such as Netflix. Bowling anyone? Or maybe you prefer car racing. Wii began life as a game console for your TV and has many neat human interfaces available for the games, such as a steering wheel. Maybe you already have Wii if there are youngsters in the house. Gaming with a Wii is another way to put the TV to use, but be warned: you might get some exercise action in the process.

TiVo

TiVo is intended to work with a cable TV service provider as a digital video recorder (DVR). You plug a provided card into the TV to access channels loaded into the card—TiVo is intended to replace the digital box from the provider. You can use TiVo and connect it to an OTR antenna also. After buying the TiVo service, you are also expected to pay a fee for added services such as TV guides. Having a TiVo lets you download video over the Internet from providers such as YouTube, Netflix, and Amazon Instant Video, to name a few. Currently TiVo will not work with satellite TV, but that might change. I doubt that TiVo would ever embrace FTA satellite. It's a great device, but if you shop from the dollar menu, you might not appreciate TiVo's monthly fees.

ROKU

For just a little bit more money, the XD/S model (current top of the line) ROKU is noticeably more feature-rich than the two entry-level models. If you're looking for an easy-to-use way to watch movies over the Internet and do not own a computer, or consider yourself amongst the "computer challenged," ROKU should be on your list to consider as a video streaming device. Another advantage is its small size and modest price. Get the HDMI cable if your TV has an HDMI slot available for ROKU, because only a composite cable is included in the box. If you travel in an RV for long periods or camp just a few weekends and do not take along a computer, the ROKU is portable and can use free campground WI-FI connections to get video content. You might have to pay to use some of the Internet entertainment services available via ROKU, such as your monthly Netflix subscription, but ROKU currently is a one-off purchase with no monthly fees.

Xbox 360

Microsoft Xbox 360 is a game console with built-in WI-FI that will also connect to the Internet for downloading video. An Ethernet port and HDMI interface is included. It also has an optical digital audio output port for connecting to a digital sound receiver. To play games with others on-line requires payment of a fee: The service is called Xbox 360 Live Gold and can be purchased for 12 months.

PS3

Sony's entry into the console market is called the PS3, short for PlayStation 3. PS3 is a gaming console that also works with Netflix. It plays Blu-ray Disc, has WI-FI built in, and also has its own gaming membership for on-line gaming on the Internet. The basic network membership is free, but an enhanced version called PlayStation Plus Network is available for a fee.

NOTE *The selections listed here are not inclusive of all the manufacturers making devices that support video streaming over the Internet. The marketplace is always shifting products and models. Those listed here are, if nothing else, some of the more popular devices and manu-*

facturers. Other devices and manufacturers include Boxee Box, Phillips, Samsung, Sharp, Seagate, and Western Digital. I suppose the one thing you can count on is that if you procrastinate long enough, you'll have more choices. If you are a risk-taker, you might want to buy any one of these devices used. Just be sure to verify that the used model you are buying contains the features you need. Consumers who want to move to the latest and greatest model of these devices often sell the old ones to help fund their desire for staying current.

On-line Sources for TV, Movies, Entertainment, and Educational Video

Following are just a few of the too many to count sources for streaming video on the Internet. In the early days, streaming video was the exclusive domain of the computer savvy in the general population. With additional devices and set top boxes that simplify access to many popular Internet sites, the process is less complicated and helps those who might have a mild case of computer-phobia. It is nice to consider your choice of tools—keyboard or remote. In any case, check some of these popular video streaming sites to see if they will provide value-added viewing and enjoyment for you and your home viewing audience.

Netflix

- **URL** www.netflix.com
- **Basic subscription fee** Under $10 per month
- **What you get** Recent movies releases, movie classics, and TV episodes numbering in the thousands.
- **What you need** Valid e-mail address and a Visa, MasterCard, or Discover credit card or a PayPal account.
- **Comment** The most popular way to get movies over the Internet.

YouTube

- **URL** www.youtube.com
- **Basic subscription fee** Free
- **What you get** Access to amateur video clips sent in by users; ability to upload your own homemade videos (you cannot upload copyrighted material).
- **What you need** Valid e-mail address, location, postal code, date of birth, and gender, and agree to abide by terms of use policy.
- **Comment** This site contains hundreds of thousands of video clips of every variety of topics you can imagine and some you can't imagine. You'll find some sweet, cute, funny, and amusing clips on this site. Site also contains material not suitable for underage viewers.

Amazon Instant Video Prime

- **URL** www.amazon.com; set the search dropdown menu to Amazon Instant Video. Or use full URL: www.amazon.com/Instant-Video/b?ie=UTF8&node=16261631.
- **Basic subscription fee** Under $100 per year
- **What you get** Ability to stream any of 5000 movies and television shows at no additional cost, a break on shipping fees, and movies to rent for a fee: typically $3.99.
- **What you need** Valid e-mail address, Amazon user account, a Visa, MasterCard, Discover, Diners Club, JCB, or American Express credit card; name, mailing address, and phone number.
- **Comment** This portion of Amazon.com is similar to Netflix and has hard copy movies for sale as well.

Pandora

- **URL** www.pandora.com
- **Basic subscription fee** Basic is free.
- **What you get** Streaming music suited to your taste in music.
- **What you need** Valid e-mail address, birth year, ZIP code, and gender.
- **Comment** This site streams music based on your preference for given artists or groups. If you are watching on screen you can give the song a thumbs up or a thumbs down. As the songs play, you are presented with an opportunity to buy the song or CD on-line. It is interposed with an occasional audio advertisement.

hulu and huluPLUS

- **URL** www.hulu.com
- **Basic subscription fee** hulu is free and advertising supported but will work only on computers in standard video. The PLUS version's fee is under $10 per month and will work on other IP devices. Check the current list for compatibility.
- **What you get** hulu: recent episodes of some TV shows, full seasons from hundreds of shows, content from ABC, FOX, and NBC. huluPLUS adds 720p resolution, current episodes, full series runs, content from ABC, FOX, NBC, and others.
- **What you need** hulu: Valid e-mail address, name, address, birth year, and gender. huluPLUS: a hulu account; Visa, MasterCard, Discover, or American Express credit card; and agree to the terms of use.
- **Comment** These two sites are all about TV show viewing. If you are not getting the networks you want because of the limited stations in your area,

this site is a good way to keep up with shows on major networks. If you use it often, the subscription version could be well worth the expense to see those shows in HD or use a device other than a computer. Check the list of compatible devices before signing up. hulu tries to limit the audience to U.S. portions of the Internet, probably because "used" TV shows here might be "first run" in syndication overseas.

As stated earlier, this listing was just the primer on some of the more popular web sites that will stream free and low-fee video, TV, and movies to your Internet-enabled computer or device. It is best to use a computer at least on the initial searches—after all, the Internet was designed as a worldwide communication system for computers of all kinds. Computers have always been able to handle working for CD, DVD and now Blu-ray Disc players, so it's no big surprise that as time goes on, many other devices are becoming more like computers. However, it is not fair to say that all of these devices are "just like" their computer ancestors in features or performance. Computers are still best for many purposes, and surfing the Web is on top of that list. Alternative devices work well for their intended purposes with compatible web sites, but if you have a computer, it's probably a good idea to keep it. If you are totally new to all of this and you can afford only one device, it's not a bad idea to start with a PC or laptop—but then again, they are not perfect for every taste and tolerance; hence the rise of the alternative devices mostly operated by touch buttons on a remote.

TV Programs, News, and Special Reports over the Internet

All major television networks have a web presence and stream full and partial program episodes, special reports, news, and weather happenings. Local stations usually have a web presence as well. When you're ready to view the web sites, you will find news, weather, and more video clips and programs than you will ever have time to watch. Some of the web sites to visit first are listed here. Look for the link to watch full program episodes. Some sites have nicely designed tabs that help you easily find the videos offered.

- www.pbs.com
- www.abc.com
- www.nbc.com
- www.cbs.com
- www.fox.com
- www.cnn.com
- www.history.com
- www.travelchannel.com
- www.biography.com

Keep in mind that the Internet is the largest library in the world, and its content is growing daily. The Library of Alexandria and Library of Congress pale in comparison. It would be impossible to list all the places where you can find mainstream, odd, interesting, funny, old, or new videos, TV shows, and movies. Many of these sources are free, some are supported by advertising fees, and others offer pay-for content.

Basic Configuration

To maintain perspective and as a final thought for this chapter, let's look at an economical configuration for viewing FTA TV and adding movie programming over the Internet while staying on a tight budget:

- TV with 720p display and multiple video inputs
- FTA satellite receiver with motorized 90 cm (36 inch) dish antenna connected to your TV that also uses local free OTA channels from an antenna
- One combination VHS/DVD player
- Basic Blu-ray Disc player connected to second HDMI port
- Computer or ROKU connected to 3 Mbps broadband connection or cellular mobile hotspot for content searching and downloading video
- Netflix on-line service fee membership
- Free hulu.com account
- Free Pandora.com account

As of this writing, the monthly cost for this basic configuration will run less than $10 per month for Netflix and approximately another $30 for an Internet service. You pay nothing to receive channels on the FTA setup. With this system, you will have hundreds of choices for watching television every day of the week and the full utility offered by the Internet, all for possibly less than $1.25 per day. Dare to compare that to your current cable, satellite, or bundled phone service bill. ("Bundle" is a company code word for taking money out of your pocket for ghost services or features you probably don't have time for or may not use anyway.) Maybe it is high time to cut the cord, chuck the cable, and dump the (fee) dish. It is your money only as long as it stays in your wallet.

Remember that you do not need any of these devices or an Internet connection to get and enjoy FTA channels on your TV. But you can add these as alternative ways to view movies and video if you think that the value proposition is there to pay the monthly fees for Internet and content providers.

CHAPTER 15

Putting It All Together for Your Home Theater Experience

Even though most of us can enjoy programming and movies from the comfort of our own homes, we still enjoy going to the theater to see current run and sometimes somewhat dated movies. The lure of the overpriced candy, popcorn, and sodas is probably not the reason many still find the experience of a trip downtown to the theater enjoyable. The very big screen and the engineered Dolby sound systems probably have much more to do with why people enjoy the full theater experience. Most of us cannot afford the screen size of a movie theater, but we can opt for the largest screen our budgets will support when planning and designing for our own home theater experience.

Having a fully equipped home theater area can provide economical entertainment opportunities for every viewer in the household. Beyond pure entertainment, a home theater system's true value added is in terms of the content that is chosen from programming sources including FTA, OTA, DVD, VCR, Blu-ray, and CD. Younger viewers can be shown math and reading courses. Teens can move to interactive video games or dance to music videos. You can enjoy audio books, foreign language courses, and educational programming from college sources, cooking classes—you name it. In this case the sky isn't the limit: with FTA, the sky is the source.

Improving your sound system to make it more theater-like will also go a long way toward increasing enjoyment of TV programming and watching videos and movies at home. Upgrading the sound system beyond what most TVs offer can easily be done on a modest budget. Providing theater-like sound is truly the one element that makes for a home "theater" environment. Comfortable seating helps a lot, too. High quality sound is the cornerstone of creating a riveting home theater experience.

Whether you are watching an adventure movie, a thundering action film, or a romantic island scene, stunning video coupled with lifelike sound effects can make you feel as if you're a part of the scene, experiencing the events in person. Even modest improvements in sound reproduction can enhance the experience of watching great videos or satellite TV programming.

Improving sound does not equal just making sound loud. When a bell is struck and you are nearby, for example, you hear the initial sound strike note and the overtone vibrations and harmonics as the bell slowly goes completely silent. The subtle nuances of created sound include echo, direction, harmonics with other sounds, and feeling sound wave vibrations that can be nearly, but never perfectly, reproduced electronically. With an ineffectual sound system, you will hear only the bell's initial note.

When movies and TV programs are made, a lot of effort goes into capturing as much of the quintessence of the sound as the technology can achieve. DVDs, CDs, and Blu-ray Discs carry enough of this sound quality to make it worthwhile to use a quality sound system for your home theater. You could spend many thousands on equipment and become an ardent audiophile who tries to squeeze the last tinkle of sound out of every sound wave form. Getting close to sound perfection is expensive, but getting to adequate can be affordable. The goal for a home theater should be getting as near to real-world sound reproduction as your budget can support.

If there is one room in your domicile that can be exclusive to television viewing or at least the dominant activity for that space, setting up for home theater can be fairly straightforward. Seven elements are *necessary* for creating a home theater. This is the place where you will want to put your best television and bring in all the cable feeds from all the external sources for video and audio programming.

Features

Let's start with a "Feature Want List" for building a home theater system. Using this list or making your own to decide how theater-like the whole experience will be when you and your viewers settle in to watch an HD movie on FTA TV will be a fun beginning to setting up a home theater.

Necessary

- Dedicated indoor room or sheltered outdoor living space
- Large flat-screen TV
- FTA receiver
- OTA digital TV receiver/converter box (if TV itself is not digital)
- Sound processor/receiver with 5.1 surround sound
- Multiple speakers
- Comfortable seating arrangement

Nice to Have

- VCR/DVD combo player

- Blu-ray Disc player
- DVR
- FM/AM receiver
- Dolby 7.1 surround sound processor with video management
- Dimmer switches on room lighting or lamps

Indulgence

- PC linked to TV
- Game box
- High-speed Internet connection
- DVR drive or network storage

Over the Top

- Dorm-size refrigerator
- Microwave
- Pod coffee maker
- Soda fountain
- Popcorn machine

System Core

The heart of a home theater system is the sound processor. For many years, the main function of the sound processor/amplifier was to decode the digital sound channels to output those sound nuances to the appropriate speakers in a multiple speaker system. Recently manufacturers, partly in response to the arrival of the HDMI interface, have recognized and met a need for either processing or passing through video as well as sound. Many of the originating video sources might not be 720p, 1080i, or 1080p, so some of the sound processors will take the older composite and component video streams and use a process called "video scaling" to reprocess the signal to appear on the TV screen as if it actually met one of the HD video standards.

As budget-minded consumers, we need to decide how many features are necessary or affordable in a sound processor installation. The minimum system is the 2.1 (aka three-channel sound) sound systems, where the sound is output to a center/woofer, a left speaker, and a right speaker. The most common system and really the minimum for fully appreciating advanced sound is the 5.1 (aka six-channel sound). The 5.1 systems have a center/woofer speaker, left front, right front, left rear, and right rear speakers. The ultimate systems are 7.1 (eight-channel sound), where a left and right channels are added to a 5.1 system. (See Chapter 12 for more information on all-around sound systems.) Having an HDMI input on the sound processor is pretty much necessary now with the prevalence of the new HDMI standard connections. Upscale sound processors can be found that not only pass the HDMI video through to the TV but can also split the signal out to two TVs. Some sound processors will be able to manage HDMI inputs for three or four devices.

Here are some key features to have on a sound processor and why they are important:

- **Multiple sound inputs** Allows sound input from more than one source
- **Minimum 50 watts sound power per speaker channel** Sound volume
- **Input for HDMI** Connection to FTA box or Blu-ray
- **Input for optical cable** Connection for optical sound output device
- **Decoding support for Dolby digital 5.1 (minimum)** Theater sound
- **Decoding support for DTS-HD** Advanced audio format
- **Audio calibration for speaker setup** Simplifies speaker setup

The prototype that can be used for comparison of sound processors is the Pioneer VSX-1019AH-K 7-Channel Home Theater Receiver. As a 7.1 sound unit, it is a bit upscale from 5.1 systems but makes a great benchmark for feature comparisons on both more and less expensive units. When you're buying on a budget, the closer a unit's feature sets (specs) are to this unit, the more likely you will be satisfied with your purchase.

Implementation in the Environment

The rest of this chapter will take a look at an understated yet typical and effective implementation of a home theater area that includes a 5.1 digital sound system and permanently mounted speakers. In this presentation the left front, right front, right rear, and left rear (LF, RF, RR, and LR) speakers are ceiling mounted, oriented downward; this sounds great, but mounting speakers lower on the wall and using speaker pedestals, with the surround speakers at ear level, also has merit.

This home, when built, was designed to make way for a home theater system. The family/living room area is in the center of the living space both spatially and for intra-functional traffic patterns. To minimize the impact, actually near zero impact, on the rest of the functional space, the TV/home theater–equipment space was built fully recessed into an interior wall. This eliminated any need for furniture or cabinetry to support the installed equipment. It also made it possible for most of the wiring to be out of sight. Companion areas on that wall are also recessed, creating a multishelved area for displaying photos, art pieces, and memorabilia or providing extra storage for a video library. Figure 15-1 shows the shelf fully integrated into the wall structure for the flat screen TV. Below the television is another assimilated and raised shelf for the sound processor, FM/AM receiver, DVD player, and

FIGURE 15-1 Stylish built-in home theater equipment space

receivers for satellite or cable. The generous depth of the shelving provides for plenty of ventilation for the TV and equipment.

A close-up of the lower shelf in this arrangement is revealed in Figure 15-2. The shelf is elevated about 6 inches above the finished floor, making access to the equipment convenient. The sound processor and FM/AM receiver are on the left.

FIGURE 15-2 Lower-shelf equipment layout

Figure 15-3 shows the sound processor, the Digital Audio/Video Control Center, that includes the newer feature of controlling the video switching. Its most important job is still processing the sound channels. Most TVs by themselves are quite capable of switching between multiple video inputs. Budget-conscious consumers take note: You can save money by purchasing a single-purpose "sound processor" to get theater quality sound without the more expensive features. It is not impossible, however, to find a good used sound-processor–only for a reasonable price.

The rest of the equipment will only be marginally useful if it's not hooked up to decent sound speakers. In the implementation featured, the homeowner chose a combination of Bose and Sony speakers. When choosing speakers, pay attention to the wattage they will handle, the frequency response range, and the minimum distortion. Keep in mind that normal hearing in normal pressures is typically 20 to 20,000 Hz. The goal is to have a system that will cover this range of normal hearing with minimal to no distortion at normal volume levels.

FIGURE 15-3 Close-up view of sound processor and AM/FM receiver

Figure 15-4 shows the woofer in the installation. Plan to purchase a 10 to 12 inch woofer at minimum, and likewise for a subwoofer, a speaker that outputs in the frequency range of 20 to 120 Hz. Woofers respond typically from 40 to 1000 Hz. The bottom line is this: To get good base level sounds, you need a speaker that will move some air. The woofer also contributes to your "feeling" the sound waves as well as hearing them.

NOTE *Remember that sound quality has not much at all to do with high volume sound alone. High volume sound will permanently damage hearing. Keep the volume at normal levels at all times; this is about quality sound, not simply loud sound.*

The left and right front speakers were handled in the same manner by mounting them high on either side of the TV. They are about 6 feet above the floor and canted down slightly so the sound cone is focused on a seated viewing audience. Figure 15-5 shows the left front speaker; the right front speaker is mirrored on the other side in exactly the same manner. These are two of the five main sound speakers, and as such they need to have the widest possible range of frequency response. That beautiful resonating bell harmonic sound is achieved only if the speakers can strike the higher frequency sounds approaching or even exceeding 20,000 Hz.

FIGURE 15-4 Woofer speaker enclosure

Figure 15-5 Wall-mounted left front channel speaker

The right rear speaker was installed on the ceiling, about 14 feet above the floor and pointed down toward the listeners. The left rear speaker is also ceiling mounted.

Although many audiophiles would say the rear speakers are mounted far too high, the quality of the speakers and the sound processor leave little to be desired when a good movie is playing on DVD, pushing sound through this processor and its attached speakers. It might not be perfect, but it is very good.

Especially if you're on a tight budget, having a sound system just a cut above "run of the mill" is doable and will surely enhance TV viewing pleasure. Remember that a computer can double as a sound processor as long as it has the outputs, and even in a one-room studio apartment, speakers can be strategically placed in positions that make the most of a 5.1 system. The sound magic occurs when the processor can do the channel splitting and reasonable quality speakers are hooked to it and are in the right places relative to the viewers.

NOTE *Speakers can also be placed on pedestals or on book shelves. Just be careful not to create tripping hazards with the wiring in smaller spaces.*

Connection Plan

Although you can't cover every conceivable hookup possibility, you can sketch out a plan for connecting everything together. Take stock of the equipment that you will be integrating into a home theater. Check the equipment or its spec sheet or manual to determine the available connections, and then sketch up a wiring diagram.

Another feature of some sound processors should be mentioned: the capability to handle connections for two sound output zones. This would allow for high quality sound in another room in the living space, such as a home office or bedroom.

Two of the most probable situations are discussed in the following sections. The first one is an older television set coupled to a 5.1 sound processor. The second is a 7.1 processor with video management/pass-through on HDMI connected to a newer television.

5.1 Processor with No HDMI Input TV

The general wiring diagram for a 5.1 sound processor and television without an HDMI connection would look similar to Figure 15-6.

Of the three RF inputs in Figure 15-6, the one that could vary from the diagram is the line from the OTA antenna. If the TV set is old and not DTV compliant, you would insert a digital converter box in this line to receive DTV signals. All the video output connectors from the video-out jacks from all of the components are cabled to the video-in jacks on the TV. All the device audio outputs, including from the TV itself if it has audio outputs, are cabled to the sound processor, yielding the best possible sound from all sources. During operation, the TV volume control is set to zero (no volume) and the sound levels are controlled by the sound processor controls or its remote. The audio cables on older devices will simply be the RCA jack style and are typically color-coded white and red.

The situation you have might vary from this diagram. To make the most of your equipment and to get high quality sound, route all the sound through the sound processor's multiple speakers.

7.1 Processor TV with HDMI Inputs

Figure 15-7 shows a diagram for more current versions of equipment and a TV that takes advantage of technology advances. The diagram shows an audio connection from the TV to the sound processor, either through an HDMI cable or through another audio output source on the TV. This is important to include because it permits the sound output from OTA channel programming to utilize the sound processor's multiple speaker system and high quality sound processing capacity.

Invest some time to ensure that you are taking advantage of the best sound performance possible from each device. If you already have a system with a sound processor, use these diagrams as a general guide to determine whether all the connections for your system are correct—that all device sound outputs are routed through the processor for best performance. As a rule, if the device has HDMI, use it. If it has no HDMI but has optical or digital coax sound outputs, use them. If it

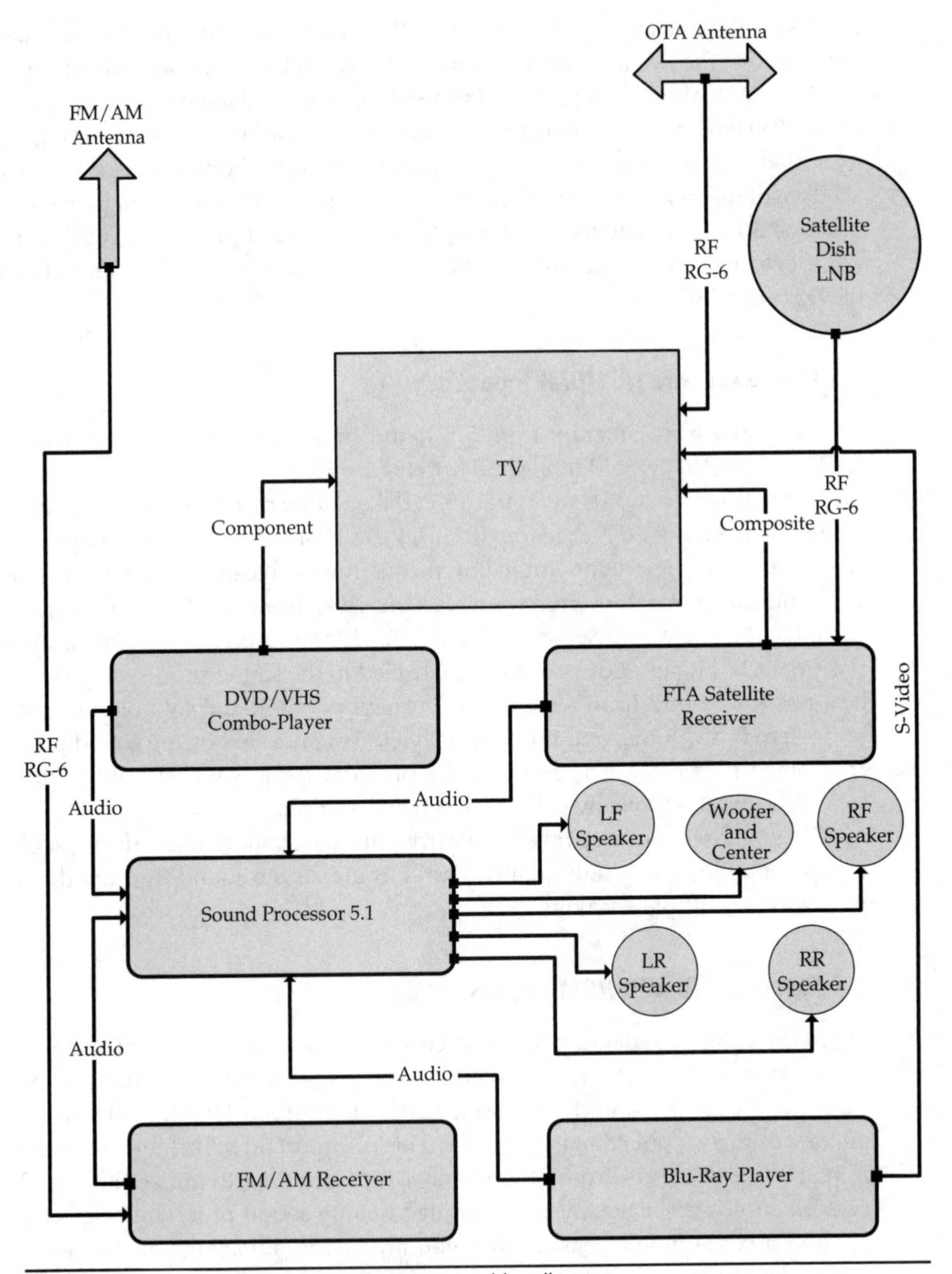

FIGURE 15-6 Older TV and 5.1 sound processor wiring diagram

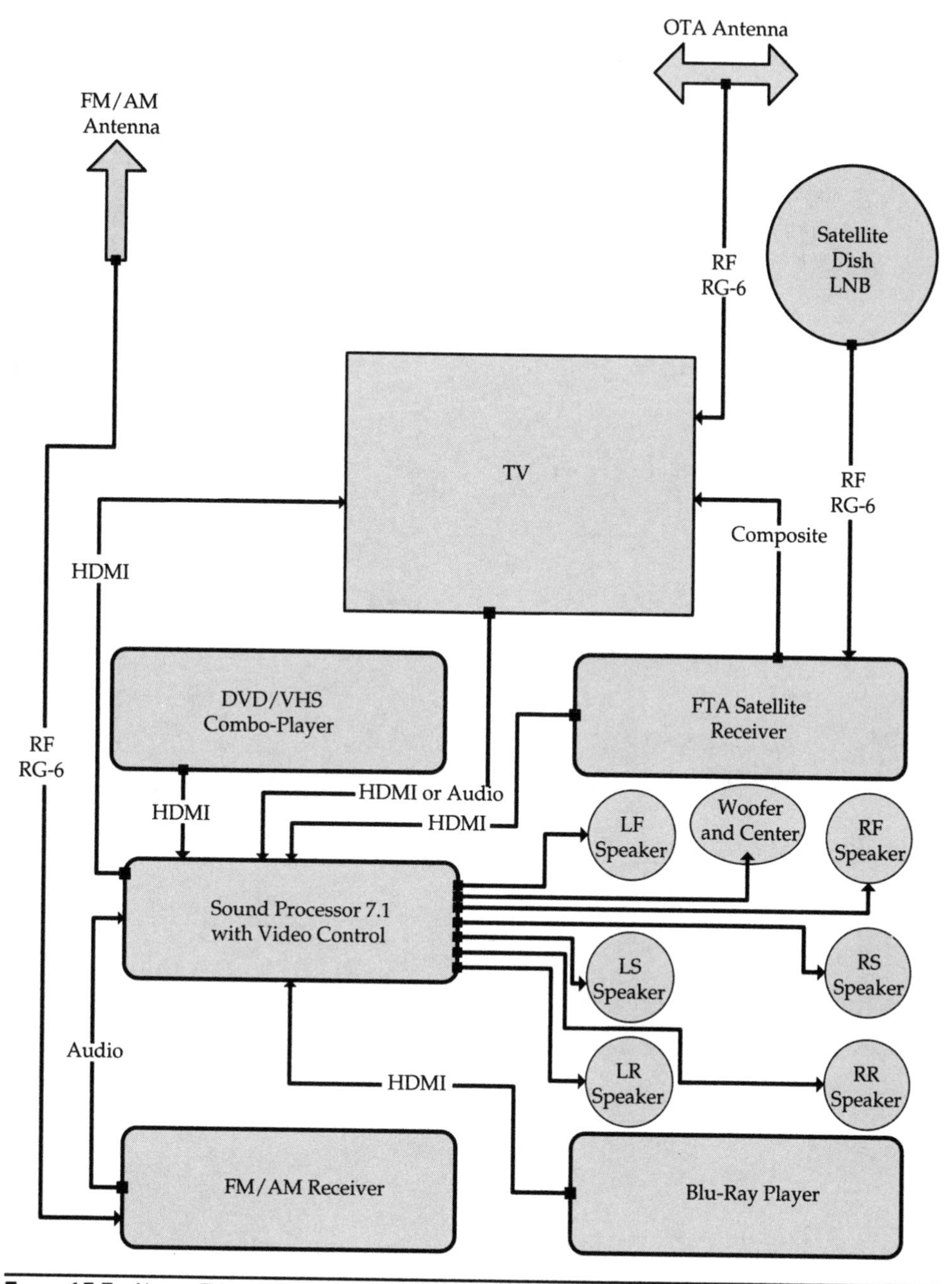

Figure 15-7 Newer TV and 7.1 sound processor wiring diagram

has no HDMI and no digital coax output and has only L/R RCA connectors, use them to connect to the sound processor.

NOTE *Connecting a computer to a large-screen television is typically done using the VGA connector on the TV and a computer video cable. To connect the computer output, use the digital coax output from the sound card to the sound processor if the computer is so equipped. The computer is not shown in the diagram; however, connection is simple when you have the correct VGA cable, usually requiring male ends on both ends of the cable.*

CHAPTER 16

Taking It with You: Mobile, RV, and Remote FTA Installations

Going mobile or remote with an FTA or pay-for services satellite system is different from doing so at home only with regard to how it will be powered and the number of times it will have to be set up to accommodate new locations. All the necessary components are readily available for installation, and setup for the first time makes for a great weekend DIY project.

To take your FTA viewing mobile or to set up at remote locations temporarily or permanently involves giving some serious thought to frugality in scale, power utilization, and setup time. The three most probable scenarios for away-from-home viewing of FTA signals are as follows:

- **Tailgating** Going mobile in a car, SUV, or truck
- **RV** Using FTA in a motor home or travel trailer
- **Remote** Installation at a cabin or camp

Not having to pay for TV programming is a substantial part of the appeal of using FTA, particularly for those on-again, off-again use scenarios such as during vacation getaways or on camping trips. Those enjoying a more mobile lifestyle, such as full-time RV'ers on a budget, might consider FTA programming as an essential supplement to whatever OTA channel programming you can receive as you caravan from place to place. Whatever the reason for wanting to bring FTA along when you are away from home, it is relatively easy and inexpensive to take it with you.

Each of these away-from-home possibilities presents similar challenges for setting up for receiving FTA signals while you're off the normal power grid. The simplest of on-the-go satellite TV systems will include a small dish, an FTA receiver, some length of cable, and a TV. Of course, the equipment will need some reliable power source as well, so we'll take a look at some options for powering up a remote FTA system. A final requirement will involve some advanced planning unless you have a very reliable mobile Internet access method. Planning will also involve finding the look angles for the dish alignment for the places from which you want to view FTA signals along the way or at the destination of your travel route.

Scale or Portability

Space for a mobile system will be more of a consideration than weight because the necessary components do not weigh much. Small televisions are readily available, from 7 inches and up, with 19-inch screens being a fairly common size now for "portable" TVs. In the final analysis, as far as going bigger than a 19-inch TV screen, I guess the mantra would be, "if you can carry it and power it, it's portable." How big is too big is something for you to decide. The same quality considerations that apply to selecting an in-home system—viewing distance, viewing angles, number of connections needed, and so forth—also apply to the choice of portable televisions.

Not much can be done as far as economizing on dish size. To receive FTA signals reasonably well, you will need at least a 75 cm (30-inch) dish. Smaller dishes, usually 18 inches on RVs such as the one shown in Figure 16-1, will not perform well when trying to receive FTA satellite signals. The RV industry assumes that you will be using Bell, Dish Network, or DIRECTV pay-for programming options, so the preinstalled dishes are small and typically have circular low noise blocks (LNBs). You can get a bracket to fit a linear LNB installed on the smaller dish, but only very close satellite signals will be captured and in fair weather only. The ability to receive FTA signals on a smaller dish fixed on a trailer or RV is further complicated by the fact that many locations will require skewing the LNB even for nearby satellites. My recommendation is to leave the preinstalled system in place and unused (unless you want to have both paid and FTA together) and to carry a 75 cm or larger dish mounted on a tripod for mobile FTA reception instead. Having the portable tripod also allows you to set up to avoid overhead tree branches that would dramatically interfere with the signal strength. Having the tripod provides much more location arrangement flexibility for setting up the dish antenna.

Fortunately, FTA receivers and other added components are not very large and weigh very little. Secure mounting and safe storage becomes more of an issue when travelling with portable electronic equipment. Keeping cords and cables under wraps so that safety is maintained and unsightly messes are avoided is another challenge. Each situation for portable use will be somewhat unique. Simply take the time and steps necessary to avoid damage to the equipment and maintain safety during its setup and operation.

Figure 16-1 Factory-installed RV satellite dish

Power Consumption

You don't need to have all Energy Star–rated components, but it is good to know how much power your on-the-go system will need for the hours of intended use. When using only 120-volt, AC-powered components, simply add up the wattage ratings of the equipment. If the equipment states only voltage and amperage for power needs, multiply the amperage times the voltage to obtain the wattage. For example, a 2.3-amp, 120-volt rated power brick is using 276 watts (2.3 × 120). Simply add up the power requirements using watts as the measure for all of the equipment you intend to use.

If you are purchasing an inverter, divide the total power needed in watts by 0.8 to size the inverter. If the total for AC-powered equipment is going to consume 480 watts while in operation, for example, you'd need to use a minimum of a 600 watt (480 watts / 0.8 = 600 watts) inverter. To calculate how long an inverter will run on an ampere-hour rated battery, divide the 600 watt example inverter unit's output in watts by 12 volts to find maximum supply amps (600 watts / 12 volts = 50 amps). A 100 ampere-hour rated battery will be able to carry this load for only 2 hours without recharge. Simply divide the rated 10-hour battery ampere-hour capacity by the maximum continuous load. Once the battery's supplied voltage goes below 10.5 volts, most inverters will shut down to prevent damage. Deep cycle batteries, not automotive starting batteries, should be used for remote power systems.

When calculating run times for a situation in which you are using all 12-volt powered equipment, simply add up the amps being used, and divide the number of amps into the ampere-hour rating of the battery (or batteries). For example, if the TV draws 7 amps and the FTA receiver draws 3 amps, that same 100 amp-hour rated battery will run them for 10 hours before needing a recharge.

One common way to recharge mobile power supply batteries is from the vehicle's alternator. With some vehicle manufacturers, you can use a battery isolator kit to charge both batteries this way. Check with the vehicle manufacturer to find out for sure. This isolator device looks like a brick with cooling fins and has three connectors on the top. It is wired between the alternator and the vehicle battery. The third connection is to the power supply battery. This arrangement allows the alternator to charge both batteries while the vehicle is running. It also prevents discharge of the vehicle battery from electric loads connected to the power supply battery. Isolation kits are available for 12-volt systems in 70- and 120-amp charge rates. The other option is to use a split-charge relay that provides similar functionality.

Setup Time

Rule number one for remote setup is to remember to bring the compass and the wrench to adjust dish elevation. If you are moving the dish frequently and want the setup times minimized, you could replace the hex nuts that adjust dish elevation on the butterfly flange with wing nuts. Hand-tightened wing nuts will work OK if the dish is not exposed to really high winds.

Setup time will be also reduced if you have taken the time to find the dish aiming criteria before arrival. Trial and error is not a very efficient method for setting up a remote dish. Setting up whatever power source should also be contrived to make setup easy and quick.

Finally, check your "going mobile kit" to ensure you have all of the components, cables, power strips, and connectors needed to assemble the system.

Choosing On-the-Go FTA Components

Moving the entire FTA system or just some of it from home is an option for going mobile. That may not be the best choice when one considers some of the reasons mobile equipment would have different features. Usually lightweight and small size are on the top of the list for mobile systems while bigger and elaborate win out for home use. The next few sections will highlight some of the considerations that apply to mobile equipment.

Mobile and Remote Power Supplies

The most important consideration for on-the-go equipment involves deciding how they will be powered during use. Household equipment in the United States is powered with 120-volt, 60 cycle, AC current and in Europe it is 230-volt, 50 cycle, AC. Frequently, newer electronic devices will have input voltage ranges for both markets. For example, you might have noticed a switch on the back of your desktop PC to switch between the two voltages. Other devices can have power supplies where the output is regulated that will work when plugged into either voltage without the need to throw a switch. Look for information on the equipment you will use on the road to see if it states a range of voltage such as 120–240 volt AC, 50/60 Hz.

One choice for providing on-the-go power is to use a gasoline- or diesel-powered generator that will provide the power and voltage needed. The other is to use an inverter that modifies an input voltage from 12- or 24-volt DC batteries into 120-volt, AC household current. Due to the hassles involved with operation of a generator, the best choice might be to use the battery/inverter option. Another option is to choose equipment that will run directly from a 12-volt DC battery source.

If the on-the-go system will be hooked up to the campground's power sources, always use a name brand surge (suppressor) protector on all the electronic equipment with sufficient ratings for protecting the connected devices. Before trusting the three-prong grounded 20 amp outlets at campgrounds, use the three-light testers shown in Chapter 2, Figure 2-25, to verify that the service ground for the outlet is connected. The surge suppressor is useless if not connected to a properly grounded outlet.

When connected to camp power sources that are not properly grounded, any level of protection from lightning strikes is also negligible. This can be extremely dangerous in open, flat areas prone to lightning strikes. If all the equipment is on board a vehicle whose only contact with the ground is rubber tires, the risk from lightning is minimized, but as soon as any contact is made with the earth via a camp power plug, leveling jacks, or a satellite dish placed on the ground, the vehicle or trailer is again in the circuit for potential lightning strikes. Having a proper ground on this power connection usually affords the only shunt-to-ground path for lightning strikes that might hit on or near the vehicle or trailer or the campground's primary or secondary power lines.

NOTE *While on the topic of safety, any connection to 120-volt or higher AC power for devices used out of doors must be connected to a ground-fault interrupter circuit (GFIC) outlet.*

TV

A few manufacturers offer 12-volt powered televisions. One example is the Jensen 19-inch LCD TV with built-in DVD player, Model FPE1909DVDC. This unit is powered either directly at 12 volts DC or can be run on house power from the included power supply. It also includes a built-in DVD player. A 19-inch screen is a nice size and can be used in the bedroom when you are at home or placed on the picnic table at a favorite campground. With the dual power choice (12- or 120-volt brick) selection, it can be used from camp power, house power, or battery power. Use this unit as a benchmark unit for comparing 12-volt television sets from other manufacturers.

When connection to 12-volt battery power is not required, simply review Chapter 10 for primary considerations for selecting any TV while keeping your unique criteria for portability in mind.

Dish

A suitable satellite dish antenna for mobile or remote use is any one of the 75 cm (30 inch) models coupled with a tripod mounting stand. A 90 cm dish such as the one featured in Chapter 4 will provide better reception over a wider range of weather conditions and will pick up more of the right and left horizon satellites, but it is hard to call 90 cm or 1.2 meter dishes portable because of the bulk of the dimensions. Rarely will you find a car trunk or even a motor home storage compartment that is big enough for a dish of either large size.

Cable

The quality and sensitivity of the receiver, the quality of the RG-6 used, and the efficiency of the LNB will be the ultimate and final arbiter of how long a cable can be for either household or on-the-go installations. The signal quality suffers first from two variables—the resistance in the cable and degradation from attenuation within the cable. Both resistance and attenuation change with cable distance. The third factor is the number of connections, with each connection adding to signal loss. So the rule of thumb is threefold:

- The cable should be only as long as necessary.
- The down-lead cable should never be longer than 200 feet.
- A 90 to 100 foot cable (nominal 30 meters) will usually work out OK.

Your portable FTA to-go kit should include one 50 foot (15 meter) cable, which will be the preferred cable to use (if you can set up the dish away from trees with a cable that short). The backup cable for use only when necessary to avoid obstacles should be about 90 feet (30 meters). An in-line connector can be used in extreme circumstances; you can use them together for 140 feet. Trying to go beyond 140 feet usually does not work out very well. If longer runs are necessary, so might be a 1.2 meter dish and/or a higher efficiency LNB, and/or an in-line amplifier—or even all three together.

FTA Receiver

Any of the FTA receivers can go portable as long as you can supply 120-volt AC power to it from camp power or your own tote-along power supply. For those who want to power the whole system from 12 volts, a handful of manufacturers are making FTA receiver units for RV (caravan) or mobile use. Two such 12-volt units are the Comag SL30/12 and the Dreambox DM 800 HD. The Dreambox also comes with a power brick rated as follows: Input 110–240 volts AC, 50–60 Hz, 0.6 amps with an output rating of 3 amps at 12 volts. It's a good choice for both home and mobile use. Use either of these two units to make feature-for-feature benchmark comparisons for 12-volt powered units from other manufacturers.

Review Chapter 6 for the other criteria important to selecting any FTA receiver.

OTA Receiver and Antenna

If you don't have a digital OTA as a feature in your television, you can purchase a DTV converter to use with an older analog TV. A few of the 12-volt DC/120-volt AC DTV-to-analog converter boxes are still available. One such model is the Airlink ATVC102 Digital TV Converter Box, which can be connected directly to 12-volt DC or used with the 120-volt AC power brick. Choose one of the antenna options for OTA outlined in Chapter 4 that is suitable for portability with sufficient quality to receive nearby stations. Chapter 9 discusses how to find stations from any location in the United States.

Rapid Setup

Rapid setup at remote locations comprises three fundamentals, and the first is having the power supply choice ready to go immediately. You can have the power supply hard-wired into the vehicle or arrange and set up the components to be quickly connected once you are on location.

Fast setup also includes being ready with the aiming information so you can point the dish to the desired satellite after arrival. If your Internet access does not go where you go, that simply means doing a map recon of the locations you will visit to find the location coordinates and doing the lookup tables for aiming information before you leave for the destination.

Finally, having a method of rapid setup for the dish itself will reduce setup time. Using a tripod mounting is one of the most efficient methods for a portable dish. The tripod legs will fold up and store easily and set up quickly. It is simply a matter of making sure the mounting pipe mast is vertical by shimming the tripod legs. Once the legs are in place, you can drive a metal stake in two of the legs to hold them in place. I have also seen sandbags used to secure the tripod legs.

FIGURE 16-2 Innovative campground dish mounting solution

Figure 16-2 shows an inventive solution for setting up a portable dish. Most campgrounds have wood picnic tables. This RV owner simply put the dish mounting flange on the picnic table and secured it with two generously sized C clamps. It is easy to visualize this method working in a number of situations. At a tailgate party, for example, you might want to secure the dish to the lip of a pickup truck bed using thin wood shims to protect the truck's paint.

Installing Mobile or Remote Electrical Power Sources

Unlike an installation at home, where at least in theory the supply of electric power is limitless, when you're going mobile, this is typically not the case. You will want to provide 120-volt AC sine wave power to be able to use your system in areas where utility power is not available. Many destination campgrounds and RV parks will have power, but if you camp, tailgate, or park in remote areas, you'll need to bring along your own power system. Sure, you could use a generator, and some very nice portable generators are available, but they come with the hassles of supplying fuel and they are noisy—often noisy enough to make the TV viewing experience less than ideal.

Two prevailing possibilities exist for powering up FTA with an inverter. The best choice to power electronic equipment, even though it is more expensive watt for watt, is to use a pure sine wave inverter instead of a sawtooth or modified sine wave (MSW) inverter. True sine wave (TSW) power is what you have at home from the local power company and what the electronic equipment is essentially designed to be powered by. By selecting a pure sine wave power inverter, you can use the most sensitive electronic equipment without fear of damaging the equipment.

Most recently manufactured RVs and travel trailers are equipped with some features to include television viewing as a part of the travel and camping experience. Prewired connections and switchboxes for satellite systems and always available power are not always included in RVs.

Adding satellite to an RV or trailer is not a formidable task and requires only basic hand tools to achieve. The same installation processes can be used to install an FTA receiver or equipment from a fee for services provider such as Dish or DIRECTV. With the fee-based services system, the difference will be the receiver itself and the polarization of the LNB.

You can provide remote or mobile power by choosing one of two styles of inverters. One style is a group of typically lower power rated units that rely on recharging the inverter's source battery from an external means such as the vehicle's own alternator. They are essentially point-of-use inverters and usually power up one or two specific 120-volt loads such as the TV and FTA receiver. One such unit is the PROwatt SW series by Xantrex, a leader in portable power. Figure 16-3 shows the PROwatt SW

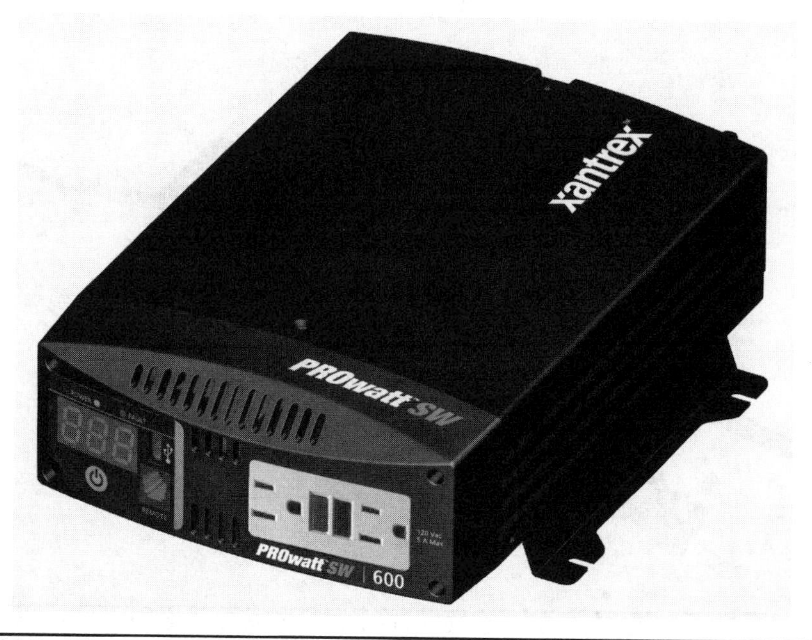

FIGURE 16-3 PROwatt SW 600 (Photo courtesy of Xantrex. Xantrex and PROwatt are trademarks of Schneider Electric Services International, registered in the U.S. and other countries.)

600 with a GFIC-rated dual outlet on the face of the unit—perfect for plugging in a TV and an FTA receiver at remote locations. This unit requires hard-wiring to the 12-volt source and should be used with a 150-amp DC in-line fuse.

Xantrex also manufactures convenient point-of-use automatic transfer switches for the SW series inverters, as shown in Figure 16-4. The part number to match the SW 600 is 806-1206. The rating of the transfer switch must match the power capacity of the inverter, so they should be ordered as a matched set. One of three cords on the transfer switch is to the equipment load, the second connects to the utility power or possibly the breaker box, and the third cord simply plugs into the outlet on the face of the inverter. When "park" AC power is not available, the inverter takes over the load. The inverter also includes USB (Type A) receptacle port for recharging or powering smaller electronics such as cell phones or WI-FI hotspots for Internet connections over the cell towers.

Figure 16-5 is an optional remote panel for controlling the on/off status of an SW series inverter from a convenient location, up to 25 feet away from the inverter.

The overview wiring diagram for connecting any point of use inverter and auxiliary battery to power it and the FTA equipment in a vehicle is shown in Figure 16-6. The 120-volt AC supply in the diagram could be from an internal outlet in a travel trailer or motor home, a campground power outlet, or a portable AC generator. The transfer switch in the diagram automatically shifts the load to the inverter if no voltage is present from the external power source. The first surge suppressor protects the relay circuit of the transfer switch from power spikes, and the second one protects the sensitive electronic equipment regardless of its source of power.

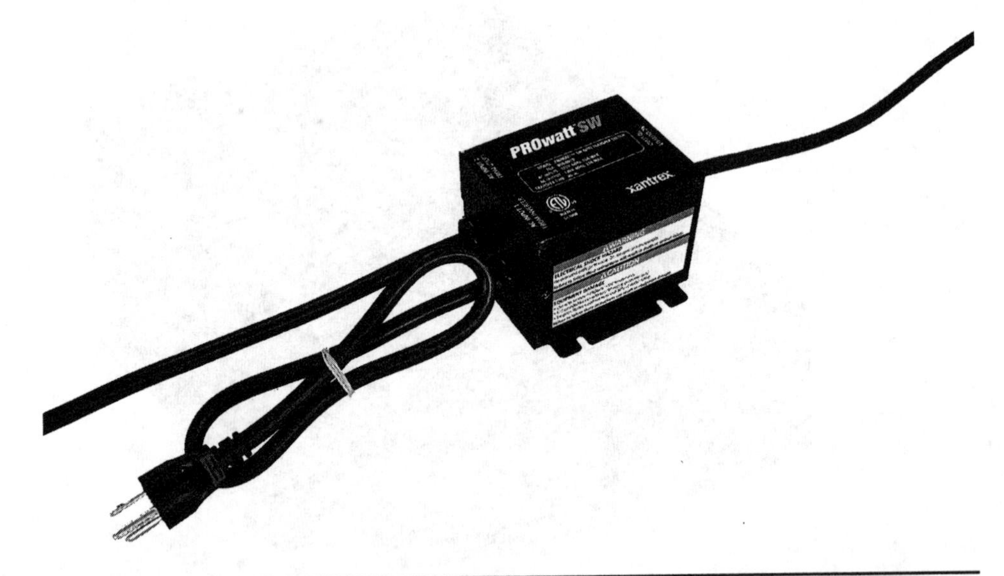

FIGURE 16-4 Matching PROwatt SW transfer switch (Photo courtesy of Xantrex. Xantrex and PROwatt are trademarks of Schneider Electric Services International, registered in the U.S. and other countries.)

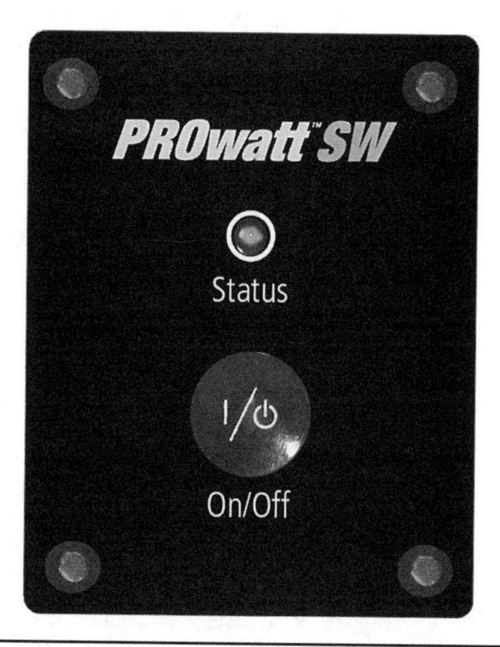

Figure 16-5 PROwatt remote control panel (Xantrex and PROwatt are trademarks of Schneider Electric Services International, registered in the U.S. and other countries.)

The alternative style of inverters that can potentially be used to support a mobile or remote FTA installation are usually physically much larger with higher AC load capacities and include built-in battery bank chargers. They are intended to supply power to multiple loads within an RV or travel trailer or even a small camp or cabin. They are more of a whole house or whole RV solution as opposed to a smaller point-of-use option. Figure 16-7 shows an example of this style, a Xantrex Freedom SW 3000. Notice the large positive and negative connection points for connecting cables to the battery bank. Remote status panels and automatic AC generator start are features that can be included with the installation. It is not impossible to accomplish a do-it-yourself installation of a unit like this; however, it would be best to seek professional help from a licensed electrician or a product-authorized dealership for installation.

Figure 16-8 shows the overall wiring scheme for installations in larger RVs. The inverter is wired to handle the connections for electronics and small appliances when the rig is operating off grid.

Installing Cable Connections in RVs or Travel Trailers

When the vehicle, RV, or travel trailer does not have an exterior connection point and prerun cables for connection of a satellite dish antenna to the outside, you might find it worthwhile to install them. The easiest way to make a hole through

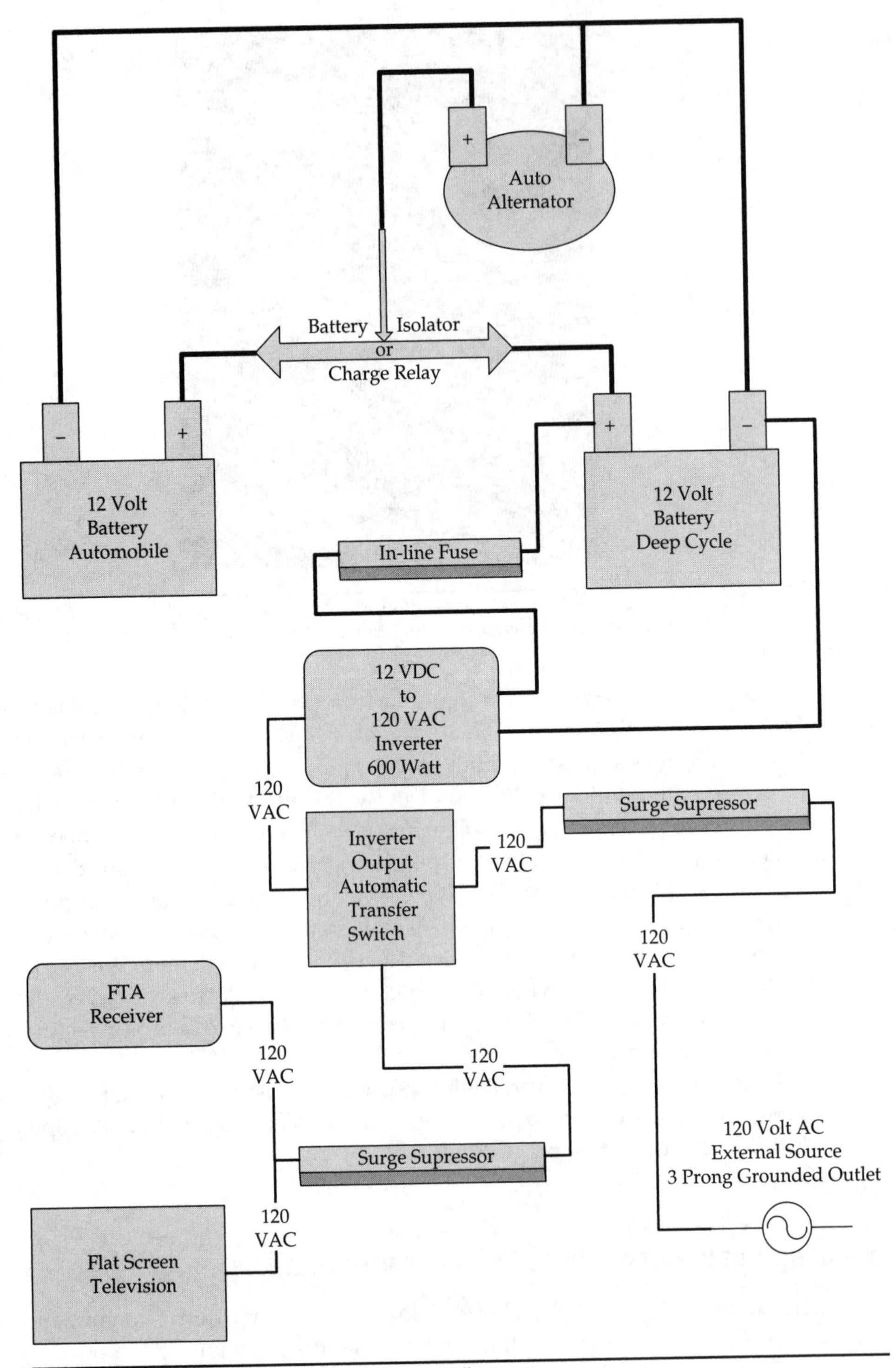

FIGURE 16-6 Point-of-use inverter vehicle wiring diagram

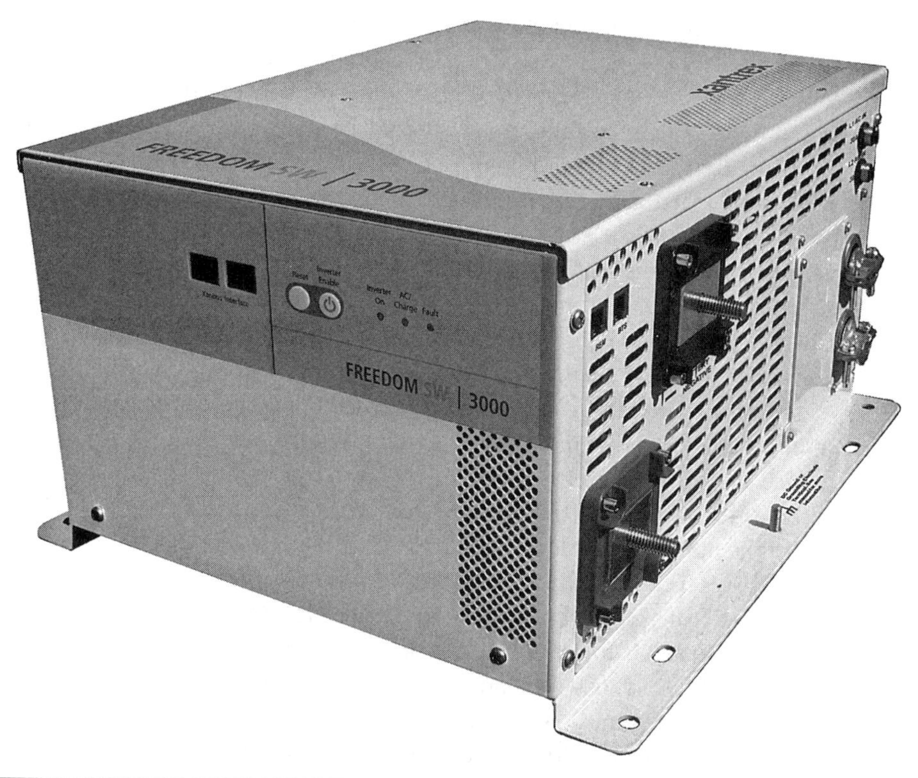

FIGURE 16-7 Freedom SW 3000 inverter/charger (Photo courtesy of Xantrex. Xantrex and PROwatt are trademarks of Schneider Electric Services International, registered in the U.S. and other countries.)

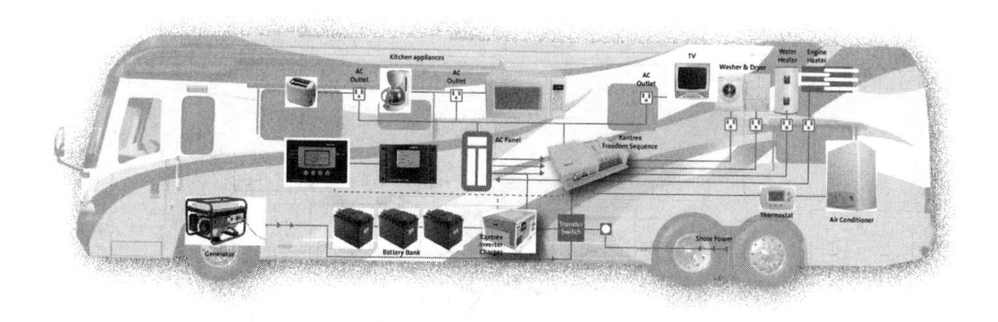

FIGURE 16-8 Wiring plan for larger motor home with inverter/charger (Diagram courtesy of Xantrex. Xantrex and PROwatt are trademarks of Schneider Electric Services International, registered in the U.S. and other countries.)

the body of a motor home is to use a 1/2-inch hole-saw. A hole-saw is sometimes referred to as a hole cutter. See Chapter 2, Figure 2-14, for a look at a fully assembled hole-saw. The unique feature of the hole-saw is that both the pilot hole and the finished hole size are cut in one operation. The hole-saw easily drills through aluminum or fiberglass exterior material, will cut through blown-in, fiberglass, or fiberglass mat insulation material, and will cut through interior wood panels or cabinetry panel walls. Use an RG-6 connector as shown in Figure 16-9 with a metal cover plate on the outside of the RV.

The longer threaded end is installed inside the vehicle. Use two interior lock washers, as shown in Figure 16-10, one on each side of the cover plate. Tighten the interior side nut very securely. After installation, it will be subjected to a lot of use by your connecting and disconnecting the cable on the outside connector as the antenna system is set up and taken down with each move.

Once the connector is securely fastened into a metal cover plate, connect the interior RG-6 cable with the F-connector to the wall plate connector and feed it through the hole in the vehicle wall from the outside. Use #6 sheet metal screws to fasten the metal cover plate to the outside of the RV if the exterior is aluminum. If the exterior is fiberglass, it is best to predrill the mounting holes to match the plate and to use pop rivets to mount the plate. To prevent water or moisture from entering through the hole, use a mastic product or plumber's putty to fill the hole just before putting the completed plate assembly in place, and use a bead of silicone caulk on the inside outer edges of the plate before pressing it against the RV siding. Once the screws or pop rivets are fully fastened tight, lay another continuous bead of caulk on the joint edges of the cover plate to match the exterior colors.

FIGURE 16-9 RG-6 wall plate connector

FIGURE 16-10 Interior end of RG-6 connector

APPENDIX A

Product Sources

Listed here are a few of the many sources for FTA components.

Antennas	www.channelmaster.com
PCT TV cards	http://hauppauge.com
Receiver manufacturers	www.AZBox.com www.captiveworks.com/products.php http://conaxtech.com/product_nano2.shtml www.dreambox.org.uk/ www.fortecstar.com www.humaxdigital.com/global/ www.matrixus.com
Small parts and cables	www.radioshack.com
RV and mobile satellite components and antennas	www.winegard.com www.campingworld.com
Power inverters	www.xantrex.com
More up-to-date information	www.allaboutfta.com

APPENDIX B

A Few of the Best Web Sites for FTA

- *Convert degrees, minutes, seconds, and decimal degrees to latitude and longitude*
  http://transition.fcc.gov/mb/audio/bickel/archive/DDDMMSS-decimal.html
- *Find azimuth with dish aiming site with moveable map pointer for receiver location*
  www.dishpointer.com
- *U.S. digital TV site*
  www.dtv.gov/whatisdtv.html
- *Canada's Transition to Digital Conversion web site*
  http://digitaltv.gc.ca/eng/1282825334983/1282825604404
- *Listing of satellites and channels by region and other data sorting, divided by regions: Africa, Asia, Europe, North America, Middle East, Pacific, South America*
  www.lyngsat.com
- *Map showing what the satellite sees*
  www.satsig.net/maps/satellite-maps-70-120-west-longitude.htm
- *StarBand satellite Internet service provider*
  http://starband.com/index.html
- *The author's site on FTA includes interesting information and links pertaining to FTA and this author's other books and articles*
  www.allaboutfta.com

Index

References to figures are in italics.